DMV Seminar
Band 8

Birkhäuser Verlag
Basel · Boston

Yum-Tong Siu

Lectures on Hermitian-Einstein Metrics for Stable Bundles and Kähler-Einstein Metrics

Delivered at the German Mathematical Society Seminar in Düsseldorf in June, 1986

1987

Birkhäuser Verlag
Basel · Boston

Author

Yum-Tong Siu
Harvard University
Department of Mathematics
Science Center
1 Oxford Street
Cambridge, MA 02138
USA

The seminar was made possible through the support of the *Stiftung Volkswagenwerk*.

Library of Congress Cataloging in Publication Data

Siu, Yum-Tong, 1943–
Lectures on Hermitian-Einstein metrics for stable bundles and Kähler-Einstein metrics.
(DMV Seminar ; Bd. 8)
Bibliography: p.
Includes index.
1. Riemannian manifolds. 2. Hermitian structures.
3. Kählerian structures. I. Deutsche Mathematiker-Vereinigung. Seminar (1986 : Düsseldorf, Germany)
II. Title. III. Series.
QA649.S5 1987 516.3'62 87-22402
ISBN 0-8176-1931-3 (U.S.)

CIP-Kurztitelaufnahme der Deutschen Bibliothek

Siu, Yum-Tong:
Lecture on Hermitian-Einstein metrics for stable bundles and Kähler-Einstein metrics : delivered at the German Math. Soc. seminar in Düsseldorf in June, 1986 / Yum-Tong Siu. – Basel ; Boston : Birkhäuser, 1987.
(DMV-Seminar ; Bd. 8)
ISBN 3-7643-1931-3 (Basel)
ISBN 0-8176-1931-3 (Boston)
NE: Deutsche Mathematiker-Vereinigung: DMV-Seminar

Printed in Germany
ISBN 3-7643-1931-3
ISBN 0-8176-1931-3

PREFACE

These notes are based on the lectures I delivered at the German Mathematical Society Seminar in Schloss Michkeln in Düsseldorf in June, 1986 on Hermitian-Einstein metrics for stable bundles and Kähler-Einstein metrics. The purpose of these notes is to present to the reader the state-of-the-art results in the simplest and the most comprehensible form using (at least from my own subjective viewpoint) the most natural approach. The presentation in these notes is reasonably self-contained and prerequisites are kept to a minimum. Most steps in the estimates are reduced as much as possible to the most basic procedures such as integration by parts and the maximum principle. When less basic procedures are used such as the Sobolev and Calderon-Zygmund inequalities and the interior Schauder estimates, references are given for the reader to look them up. A considerable amount of heuristic and intuitive discussions are included to explain why certain steps are used or certain notions introduced. The inclusion of such discussions makes the style of the presentation at some places more conversational than what is usually expected of rigorous mathemtical presentations.

For the problems of Hermitian-Einstein metrics for stable bundles and Kähler-Einstein metrics one can use either the continuity method or the heat equation method. These two methods are so very intimately related that in many cases the relationship betwen them borders on equivalence. What counts most is the *a priori* estimates. The kind of scaffolding one hangs the *a priori* estimates on, be it the continuity method or the heat equation method or even the method of minimizing sequences, is of rather minor importance when the required *a priori* estimates are available.

For variety's sake we choose the heat equation approach for the problem of Hermitian-Einstein metrics for stable bundles and choose the continuity method for the problem of Kähler-Einstein metrics. At the time these lectures were given Donaldson's heat equation method for the problem of Hermitian-Einstein metrics for stable bundles was done only for the surface case. Later he improved his method to make it work also for the general case. In these notes we present his improved version though only the surface case was lectured on in Düsseldorf. The problems for the existence and uniqueness of Hermitian-Einstein metrics for stable bundles and of Kähler-Einstein metrics for the case of negative and zero anticanonical class have been completely solved. The contributors to the original solutions and the subsequent simplications are Aubin, Bouguignon, Calabi, Buchdahl, Donaldson, Evans, M.S. Narasimhan, Seshadri, Uhlenbeck, and Yau. Individual contributions will be detailed in the sections where the material is presented.

The problem that is still open in this area concerns Kähler-Einstein metrics for the case of positive anticanonical class. In that case there are obstructions to the existence of Kähler-Einstein metrics. One obstruction is the non-reductivity of the automorphism group discovered by Matsushima and Lichnerowicz. The other obstruction is the nonvanishing of an invariant for holomorphic vector fields due to Kazdan, Warner, and Futaki. The uniqueness problem for Kähler-Einstein metrics up to biholomorphisms was recently solved by Bando and Mabuchi. The existence problem for Kähler-Einstein metrics for the case of positive anticanonical class is still very open. I briefly discuss a very minor recent existence result of mine for the case when the manifold admits a suitable finite symmetry. The applicability of this method is exceedingly limited. For surfaces it works for the Fermat cubic surface and the surface obtained by blowing up three points of the complex projective surface. It can also be applied to higher dimensional Fermat hypersurfaces. The conjecture that any compact Kähler manifold with positive anticanonical class and no nonvanishing holomorphic vector fields admits a Kähler-Einstein metric is still unsolved. Any meaningful contribution to the existence problem of Kähler-Einstein metrics for the positive anticanonical class case should make substantial use of holomorphic vector fields or their nonexistence, which unfortunately nobody knows how to do up to this point.

In these notes we do not discuss Calabi's theory of extremal Kähler metrics which are critical points of the functional of the global square norm of the curvature tensor. Neither do we discuss the applications of the existence of Kähler-Einstein metrics such as the uniqueness of complex structure on the complex projective space admitting a Kähler metric and the existence of Kähler metric on every K3 surface.

I would like to thank Professor Gerd Fischer of the University of Düsseldorf who organized and invited me to the German Mathematical Society Seminar in Düsseldorf at which these lectures were delivered and who arranged and encouraged the publication of these lecture notes. During the preparation of these lecture notes I was partially supported by a National Science Foundation grant and a Guggenheim Fellowship.

Y... ng SIU

University

...e, Massachusetts

TABLE OF CONTENTS

CHAPTER 1. THE HEAT EQUATION APPROACH TO HERMITIAN-EINSTEIN METRICS ON STABLE BUNDLES

In this chapter we discuss here the problem of the existence of Hermitian-Einstein metrics on stable bundles. The case of stable bundles over compact complex curves was done by Narasimhan and Seshadri [N-S] and alternative proof later given by Donaldson [D1]. Donaldson [D2] treated the case of stable bundles over compact algebraic surfaces by using the heat equation method. Recently a paper of Uhlenbeck and Yau [U-Y] dealt with the problem for stable bundles over a compact Kähler manifold by using the continuity method. A recent preprint of N. P. Buchdahl [Bu] considered the case of stable bundles over compact complex surfaces that admit a $\partial\bar{\partial}$-closed positive definite (1,1)-form. A most recent preprint of Donaldson [D3] carried over his treatment of the surface case to higher dimension. Simpson [Si] in his Harvard University dissertation under the direction of W. Schmid proved the existence of Hermitian-Einstein metrics for stable systems of Hodge bundles without the restriction of algebraicity.

We will treat here only the method of Donaldson and retriction our discussions to that approach. For the heat equation approach used by Donaldson, to get *a priori* estimates instead of using directly the definition of stable bundles he used the following property of stable bundles. The restriction of a stable bundle to a generic hypersurface cut out by a sufficiently high powers of an ample divisor is again stable. At the time the lectures were delivered at Düsseldorf Donaldson's preprint [D3] on the higher-dimensional case was not yet available and the lectures presented only his treatment of the surface case. The recent improvement Donaldson made on his method makes his proof more elegant and more natural. It serves no useful purpose to stick to his earlier treatment [D2] of the surface case which was originally presented in Düsseldorf. So in these notes we present his improved version which works in any dimension. We follow closely Donaldson's proof except that instead of estimating the Laplacian of the logarithm of the maximum of the eigenvalues of the Hermitian metric with respect to a fixed background Hermitian metric, we estimate the Laplacian of the logarithm of the trace. Not only is the estimation of the Laplacian of the logarithm of the trace more natural and easier, but also it is analogous to the estimation of Aubin [A2, p.120, (β)] on $\Delta' \log(m+\Delta\varphi)$ in the Kähler-Einstein case which is the same as Δ' of the logarithm of the trace of the new Kähler metric with respect to the old one.

§1. *Definition of Hermitian-Einstein Metrics.*

(1.1) Let M be a complex manifold of complex dimension m and E be a holomorphic vector bundle of $\mathbb{C}$-rank r over M. Let H be a Hermitian metric along the fibers of E. With respect a local trivialization of E the Hermitian metric H is a positive Hermitian matrix $(H_{\alpha\bar{\beta}})_{1\leq\alpha,\beta\leq r}$. We are going to use the first index α as the row index and the second index β as the column index for the matrix $(H_{\alpha\bar{\beta}})_{1\leq\alpha,\beta\leq r}$.

We introduce the concept of a *complex metric connection* for the Hermitian vector bundle E with metric H. A connection is a way of defining the partial derivative of a section so that the result is also a section. Suppose e_α, $1 \leq \alpha \leq r$, is a smooth local basis of E. Let D denote the operator of differentiating sections and we use D to mean total differentiation so that the information of partial differentiation in all directions is contained in D. The result De_α of applying D to e_α is a local E-valued 1-form on M. We express De_α in terms of the basis e_β, $1 \leq \beta \leq r$ and get $De_\alpha = A_\alpha^\beta e_\beta$, where A_α^β is a local 1-form on M and the summation convention of summing over repeated indices is used. We use A to denote the matrix $(A_\alpha^\beta)_{1\leq\alpha,\beta\leq r}$ and regard the subscript α as the row index and the superscript β as the column index of the matrix $(A_\alpha^\beta)_{1\leq\alpha,\beta\leq r}$. For convenience's sake we also call the matrix A the connection.

Since the bundle E is holomorphic, it is possible to define partial differentiation in the (0,1) direction in a natural way, *viz.* the (0,1) derivative of a local holomorphic section of E is defined to be zero and the (0,1) derivative of any smooth section is defined by expressing it in terms of a local holomorphic basis and uisng the Leibniz rule of differentiating products. A connection is said to be *complex* if its partial differentiation in the (0,1) direction is the natural one just described. When the local basis e_α $(1 \le \alpha \le r)$ is holomorphic, a connection $(A_\alpha^{\ \beta})_{1\le\alpha,\beta\le r}$ is complex if and only if the local 1-forms $A_\alpha^{\ \beta}$ are all of type (1,0). From now on unless the contrary is explicitly mentioned we deal only with complex connections.

When we have a connection, we can define the concept of a parallel section of E along a curve γ of M. Such a section is one whose partial derivative along γ vanishes. A connection is said to be *metric* if the length of any parallel section of E along any curve γ is necessarily constant along γ. By using a parallel unitary frame along γ, it is easy to see that a connection is metric if and only if

$$d\langle u,v\rangle = \langle Du,v\rangle + \langle u,Dv\rangle \tag{1.1.1}$$

for any local smooth section u and v of E, where $\langle\cdot,\cdot\rangle$ denotes the pointwise inner product defined by the metric H and the equation means that both sides give the same value when evaluated at any tangent vector of M. For $u = e_\alpha$ and $v = e_\beta$ equation (1.1.1) reads

$$dH_{\alpha\bar\beta} = A_\alpha^{\ \gamma} H_{\gamma\bar\beta} + H_{\alpha\bar\gamma}\overline{A_\beta^{\ \gamma}}$$

as one can easily verify by evaluating both sides at a tangent vector of M. In matrix notations this means that

$$dH = AH + H\bar{A}^t,$$

where $\bar{A}$ is the complex conjugate of A and the superscript t of $\bar{A}$ means the transpose of $\bar{A}$. By breaking down the equation into the (1,0) and (0,1) components, we get two equations $\partial H = AH$ and $\bar\partial H = H\bar{A}^t$, because A, being a complex connection, is a matrix of (1,0) forms. The two equations

$\partial H = AH$ and $\bar{\partial} H = H\bar{A}^t$ are equivalent, because H is a Hermitian matrix, as one can easily see by evaluating at a tangent vector of M and taking complex conjugates and transposes of the matrices. From now on unless the contrary is explicitly mentioned we deal only with metric connections.

Given any Hermitian metric H along the fibers of a holomorphic vector bundle E there exists one and only one complex metric connection A. We verify the statement by taking a local holomorphic basis e_α $(1 \leq \alpha \leq r)$. The two conditions are: (i) A is a matrix of (1,0)-forms; and (ii) $\partial H = A$. Thus $A = (\partial H)H^{-1}$ is the unique complex metric connection. Sometimes to emphasize the dependence of A on H we also denote A by A_H. In the literature sometimes the connection is written as $H^{-1}\partial H$ because of the reversal of the rôles of the row and column indices.

(1.2) Unlike the case of partial differentiation for functions, in general for sections of a vector bundle partial differentiations for different directions do not commute. The failure of the commutativity is measured by the curvature of the connection. The commutativity of partial differentiation of a function in two different directions is equivalent to the vanishing of the composite of two exterior differentiation applied to functions. Take a local smooth basis e_α $(1 \leq \alpha \leq r)$ of E. We apply D twice to e_α and get a local section $D\,D\,e_\alpha$ of $E \otimes T^*_M \otimes T^*_M$, where T^*_M is the dual of the tangent bundle T_M of M. By skew-symmetrizing $D\,D\,e_\alpha$ in the two arguments for the tangent vectors of M, we get a local E-valued 2-form $D\wedge D\,e_\alpha$. We can express $D\wedge D\,e_\alpha$ in terms of the connection A and its exterior derivative dA as follows. We use the column vector e with components e_α.

$$\begin{aligned} D\wedge D\,e &= D\wedge(A\,e) = dA\,e - A\wedge De \\ &= dA\,e - A\wedge A\,e = (dA - A\wedge A)\,e. \end{aligned}$$

We define the *curvature* to be $dA - A\wedge A$ and denote it by F_H or F_A or simply by F. It is an End(E)-valued 2-form on M, because if f is a local smooth matrix-valued function, then

$$\begin{aligned} D\wedge D(fe) &= D\wedge(df\ e + f\ De) \\ &= ddf\ e - df\wedge De + df\wedge De + fD\wedge De = f\ D\wedge De. \end{aligned}$$

When we choose a local holomorphic basis e_α $(1 \le \alpha \le r)$, we have

$$\begin{aligned} F_H &= dA - A\wedge A \\ &= d((\partial H)H^{-1}) - (\partial H)H^{-1}\wedge(\partial H)H^{-1} \\ &= (\partial + \bar\partial)((\partial H)H^{-1}) - (\partial H)H^{-1}\wedge(\partial H)H^{-1} \\ &= (\partial H)H^{-1}\wedge(\partial H)H^{-1} + \bar\partial((\partial H)H^{-1}) - (\partial H)H^{-1}\wedge(\partial H)H^{-1} \\ &= \bar\partial((\partial H)H^{-1}) = \bar\partial\ A. \end{aligned}$$

This shows that F_H is actually an End(E)-valued (1,1)-form on M and it simply equals $\bar\partial\ A$ when expressed in terms of a local holomorphic basis of E. In the literature usually the curvature tensor is defined as $\frac{i}{2\pi} F_H$ instead of F_H so that the curvature tensor is real and its integral over a curve equals the Chern number of the restriction of the bundle to the curve.

We denote by $\mathrm{Tr}\ F_H$ the trace of F so that $\mathrm{Tr}\ F_H$ is a (1,1) form on M. The (1,1)-form $\mathrm{Tr}\ F_H$ is the curvature form of the determinant bundle det E of E with the metric det H induced from the metric H of E, because with respect to a local holomorphic basis

$$\partial \log \det H = H^{\alpha\bar\beta}\ \partial\ H_{\alpha\bar\beta} = \mathrm{Tr}((\partial H)H^{-1})$$

and

$$\bar\partial((\partial(\det H))(\det H)^{-1}) = \mathrm{Tr}(\bar\partial((\partial H)H^{-1}) = \mathrm{Tr}\ F.$$

The curvature tensor F satisfies a Bianchi identity (corresponding to the second Bianchi identity for the Riemannian curvature tensor). The Bianchi identity is

$$\begin{aligned} D\wedge F &= dF + F\wedge A - A\wedge F \\ &= d(dA - A\wedge A) + (dA - A\wedge A)\wedge A - A\wedge(dA - A\wedge A) \\ &= -\,dA\wedge A + A\wedge dA + dA\wedge A - A\wedge A\wedge A - A\wedge dA + A\wedge A\wedge A = 0. \end{aligned}$$

Here D denotes the covariant differentiation of $\mathrm{End}(E)$ induced from that of E. In terms of local coordinates $\nabla_k F_\alpha{}^\beta{}_{i\bar j} + \nabla_i F_\alpha{}^\beta{}_{\bar j k} + \nabla_{\bar j} F_\alpha{}^\beta{}_{ki} = 0$. Since the curvature tensor is of type (1,1), we simply have $\nabla_k F_\alpha{}^\beta{}_{i\bar j} = \nabla_i F_\alpha{}^\beta{}_{k\bar j}$. Analogously we have also $\nabla_{\bar k} F_\alpha{}^\beta{}_{i\bar j} = \nabla_{\bar j} F_\alpha{}^\beta{}_{i\bar k}$.

(1.3) We now introduce the concept of a Hermitian-Einstein metric. Let ω be a Kähler form on M. We denote by ΛF the contraction of F with ω. More precisely, if $\omega = \sqrt{-1}\, g_{i\bar j}\, dz^i \wedge \overline{dz^j}$ and $F = (F_\alpha{}^\beta)$ with $F_\alpha{}^\beta = F_\alpha{}^\beta{}_{i\bar j}\, dz^i \wedge \overline{dz^j}$, then $(\Lambda F)_\alpha{}^\beta = g^{i\bar j} F_\alpha{}^\beta{}_{i\bar j}$, where $(g^{i\bar j})$ is the inverse trix of $(g_{i\bar j})$.

(1.3.1) *Definition*. The Hermitian metric H along the fibers of the holomorphic vector bundle E over a Kähler manifold M is *Hermitian-Einstein* if $\Lambda F_H = \gamma I$ at every point of M, where γ is a global constant and I is the identity endomorphism of E.

From this point on we assume that the Kähler manifold M is compact and the complex dimension of M is m.

(1.4) *Remarks*. (i) When one has a Hermitian metric H with $\Lambda F_H = \gamma I$ for some *pointwise* constant γ, then it is possible to make a conformal change in the metric H to get a Hermitian-Einstein metric. The reason is as follows. We construct a new metric H' which is related to the old metric by the following conformal change $H' = e^{-\varphi} H$. Let $F = F_H$ and $F' = F_{H'}$. Then $F' = F + \partial\bar\partial\varphi$ and $\Lambda F' = \Lambda F + \Delta\varphi$. Let $\hat\gamma$ be the average of γ over M. Then $\int_M (\gamma - \gamma') = 0$. (When a measure of volume form is missing from an integral, the volume fomr from the Kähler metric is being used unless the contrary is explicitly stated.) There exists φ such that $\Lambda\varphi = \gamma - \hat{\gamma}$. Then $\Lambda F' = \hat{\gamma}I$.

(ii) A closed (1,1)-form ξ is harmonic if and only if $\Lambda\xi$ = constant. For the "only if" part, we observe that $\xi\wedge\omega^{m-1}$ is harmonic, because the product of two harmonic forms is harmonic as one can check by applying to it the operators $\bar{\partial}$ and $\bar{\partial}^*$. There is only one harmonic form ω^m of top degree up to a constant factor. Hence $\xi\wedge\omega^{m-1} = c\omega^m$ which means that $\Lambda\xi$ is constant. For the "if" part, we let η be the harmonic representative of ξ in its class. Then $\xi = \eta + \partial\bar{\partial}f$ and $\Lambda\xi = \Lambda\eta + \Lambda f$. Since ξ, η are in the same class, $\int_M \xi\Lambda\omega^{m-1} = \int_M \eta\Lambda\omega^{m-1}$ which means that $\int_M \Lambda\xi = \int_M \Lambda\eta$. But both $\Lambda\xi$ and $\Lambda\eta$ are constants. So $\Lambda\xi = \Lambda\eta$ and $\Lambda f = 0$ and f is constant and $\xi = \eta$.

(iii) Let $F^0 = F - \left[\frac{1}{\text{rk } E}\text{Tr}_H F\right]I$ be the trace-free part of F. The bundle (E,H) is Hermitian-Einstein if and only if $\Lambda F^0 = 0$ and $\text{Tr}_H F$ is harmonic. For the "if" part, since $\Lambda F^0 = \Lambda F - \left[\frac{1}{\text{rk } E}\text{Tr}_H \Lambda F\right]I$, we have $\Lambda F = \gamma I$ when we set $\frac{1}{\text{rk } E}\text{Tr}_H\Lambda F = \gamma$. Moreover, the harmonicity of $\text{Tr}_H F$ implies that $\frac{1}{\text{rk } E}\text{Tr}_H\Lambda F$ is a constant. For the "only if" part, $\Lambda F = \gamma I$ implies that $\Lambda\text{Tr}_H F = \gamma \text{ rk } E$ is constant and $\text{Tr}_H F$ is harmonic. Clearly $\Lambda F = \gamma I$ implies that $\Lambda F^0 = 0$.

(iv) Suppose we have a holomorphic family of compact complex manifolds M_s $(s \in S)$ with parameter space S which is a complex manifold and suppose each member of the family is given a Kähler form ω_s which varies smoothly as a function of the variable s of S. Assume that we have a holomorphic vector bundle E_s over each member M_s of the family so that these bundles together form a holomorphic bundle over the whole family. Suppose $s_0 \in S$ and H_{s_0} is a Hermitian-Einstein metric of E_{s_0} with respect to the Kähler form ω_{s_0}. Suppose that E_{s_0} admits no global holomorphic endomorphisms over M_{s_0} other than mulitiples of the identity. Then there exist an open neighborhood U of s_0 in S and Hermitian-Einstein metrics H_s of E_s with respect to ω_s for $s \in U$ so that H_s varies smoothly in s. This one can see by using the implicit function theorem. Let ∂_s and $\bar\partial_s$ be the (1,0) and (0,1) exterior differential operator of the complex manifold M_s and Λ_s be the contraction operator with respect to the Kähler form ω_s. We consider the following equation which defines the Hermitian-Einstein property of the Hermitian metric H_s of E_s.

$$\Lambda_s(\bar\partial_s(\partial_s H_s)H_s^{-1}) = \lambda I,$$

where λ is the constant depending on the topology of E_s and is independent of s. This equation has a solution at s_0. To be able to apply the implicit function theorem to get a solution H_s for s near s_0, it suffices to show that for any smooth endomorphism L of E_{s_0} the equation $\Delta_{s_0} K = L$ can be solved for the unknown K which is a smooth endomorphism of E_{s_0}, where Δ_{s_0} is the Laplace-Beltrami operator for $\operatorname{End}(E_{s_0}, E_{s_0})$ with respect to the Kähler metric ω_{s_0} of M_{s_0} and the Hermitian metric H_{s_0} of E_{s_0}. This equation can always be solved, because the kernel of Δ_{s_0} is zero due to the nonexistence of global holomorphic endomorphisms of E_{s_0} over M_{s_0} other than homotheties.

(1.5) We now introduce stability. Let E be a holomorphic vector bundle of $\mathbb{C}$-rank r over a compact Kähler manifold M of complex dimension m. For a subbundle E' of E (or a coherent subsheaf E' of E with torsion-free quotient E/E') and rank s, by the normalized first Chern number $\mu(E')$ we mean $\frac{1}{s} c_1(E')[\omega]^{m-1}$, where $[\omega]$ denotes the cohomology class defined by the Kähler form ω. The vector bundle E is said to be *stable* with respect to the Kähler class $[\omega]$ of M if $\mu(E') < \mu(E)$ for every proper coherent subsheaf E' of E with torison-free quotient E/E'. If $\mu(E') < \mu(E)$ is replaced by $\mu(E') \leq \mu(E)$, then we say that E is *semistable*.

A stable bundle E admits no global holomorphic endomorphisms other than homotheties. For if σ is a global holomorphic endomorphism of E which is not a multiple of the identity, then at some point P of M there exists an eigenvalue τ of σ which is not equal to all the eigenvalues of σ at P. Let E' and E'' be the kernel and the image of $\sigma - \tau I$. Then $c_1(E) = c_1(E') + c_1(E'')$ and $rk(E) = rk(E') + rk(E'')$. From $\mu(E') < \mu(E)$ and $\mu(E'') < \mu(E)$ we obtain the contradiction $\mu(E) < \mu(E)$.

We prove here the following proposition due to Kobayashi [Ko] and Lübke [Lü2].

(1.6) *Proposition*. If E admits a Hermitian-Einstein metric, then E is an orthogonal direct sum of stable bundles.

Proof. First we assume that E is cannot be decomposed into a nontrivial direct sum of holomorphic subbundles which are orthogonal to each other. Let E' be a proper coherent subsheaf of E' of rank s with torsion-free quotient E/E'. At a point of M where E' is a subbundle of E, we choose a local holomorphic basis e_α $(1 \leq \alpha \leq r)$ of E so that e_α $(1 \leq \alpha \leq s)$ is a section of E'. Let $H = (H_{\alpha\bar{\beta}})$ be the Hermitian metric of E expressed in terms of this basis. Then $F = \bar{\partial}((\partial H)H^{-1}) = (\bar{\partial}\partial H)H^{-1} + \partial H\, H^{-1}\wedge\bar{\partial}H\, H^{-1}$ and $FH = \bar{\partial}\partial H + \partial H\, H^{-1}\wedge\bar{\partial}H$. The tensor FH is simply the tensor obtained by lowering the index of F and we use $F_{\alpha\bar{\beta}}$ to denote its components. So we have $F_{\alpha\bar{\beta}} = \bar{\partial}\partial H_{\alpha\bar{\beta}} + \partial H_{\alpha\bar{\gamma}} H^{\bar{\gamma}\delta}\bar{\partial}H_{\delta\bar{\beta}}$. We assume that the basis e_α $(1 \leq \alpha \leq r)$ is chosen to be unitary at the point under consideration. Then $F_{\alpha\bar{\beta}} = \bar{\partial}\partial H_{\alpha\bar{\beta}} + \Sigma_{\gamma=1}^{r}\partial H_{\alpha\bar{\gamma}}\wedge\bar{\partial}H_{\gamma\bar{\beta}}$. Likewise the curvature tensor $F'_{\alpha\bar{\beta}}$ of E' is given by $F'_{\alpha\bar{\beta}} = \bar{\partial}\partial H_{\alpha\bar{\beta}} + \Sigma_{\gamma=1}^{s}\partial H_{\alpha\bar{\gamma}}\wedge\bar{\partial}H_{\gamma\bar{\beta}}$ when α and β are between 1 and s. So $F_{\alpha\bar{\beta}} = F'_{\alpha\bar{\beta}} + \Sigma_{\gamma=s+1}^{r}\partial H_{\alpha\bar{\gamma}}\wedge\bar{\partial}H_{\gamma\bar{\beta}}$ and the $E'\otimes\overline{E'}$-valued (1,1)-form $\Sigma_{\gamma=s+1}^{r}\partial H_{\alpha\bar{\gamma}}\wedge\bar{\partial}H_{\gamma\bar{\beta}}$ is semipositive in the sense that for any s-tuple $(\xi^{\alpha})_{1\leq\alpha\leq s}$ of complex numbers $\Sigma_{1\leq\alpha,\beta\leq s}(\Sigma_{\gamma=s+1}^{r}\partial H_{\alpha\bar{\gamma}}\wedge\bar{\partial}H_{\gamma\bar{\beta}})\xi^{\alpha}\overline{\xi^{\beta}}$ is a semipositive (1,1)-form. When one contracts both curvature tensors with Kähler form ω of M, one gets $(\Lambda F)_{\alpha\bar{\beta}} = (\Lambda F')_{\alpha\bar{\beta}} + \Sigma_{\gamma=s+1}^{r}g^{i\bar{j}}\partial_i H_{\alpha\bar{\gamma}}\ \partial_{\bar{j}}H_{\gamma\bar{\beta}}$ and $\Sigma_{\gamma=s+1}^{r}g^{i\bar{j}}\partial_i H_{\alpha\bar{\gamma}}\ \partial_{\bar{j}}H_{\gamma\bar{\beta}}$ is a semipositive $s\times s$ Hermitian matrix in the two indices α and $\bar{\beta}$.

Let λ be the complex number which is equal to the average of $\frac{1}{r}\,\mathrm{Tr}(\Lambda F)$ over M. This number λ is the number that satisfies $F_{\alpha\bar{\beta}} = \lambda\ H_{\alpha\bar{\beta}}$ when the Hermitian metric H is Hermitian-Einstein. The condition that $\mu(E') < \mu(E)$ is equivalent to $\frac{1}{s}\int_M \mathrm{Tr}(\Lambda F') < \frac{1}{r}\int_M \mathrm{Tr}(\Lambda F)$. That is $\frac{1}{s}\int_M \mathrm{Tr}(\Lambda F') < \lambda\ \mathrm{Vol}(M)$. Let $\theta_{\alpha\bar{\beta}}$ be $\Sigma_{\gamma=s+1}^{r}g^{i\bar{j}}\partial_i H_{\alpha\bar{\gamma}}\ \partial_{\bar{j}}H_{\gamma\bar{\beta}}$. Then the condition can be rewritten as $\frac{1}{s}\int_M \Sigma_{\alpha=1}^{s}\ \Lambda F_{\alpha\bar{\alpha}} < \lambda\ \mathrm{Vol}(M) + \frac{1}{s}\int_M \Sigma_{\alpha=1}^{s}\ \theta_{\alpha\bar{\alpha}}$.

Since $\Lambda F_{\alpha\bar{\beta}} = \lambda H_{\alpha\bar{\beta}}$ for $1 \leq \alpha,\beta \leq s$, clearly $\frac{1}{s}\int_M \Sigma_{\alpha=1}^{s}\ \Lambda F_{\alpha\bar{\alpha}} = \lambda\ \mathrm{Vol}(M)$ and the inequality $\frac{1}{s}\int_M \Sigma_{\alpha=1}^{s}\ \Lambda F_{\alpha\bar{\alpha}} \leq \lambda\ \mathrm{Vol}(M) + \frac{1}{s}\int_M \Sigma_{\alpha=1}^{s}\ \theta_{\alpha\bar{\alpha}}$ is satisfied with equality precisely when $\partial\ H_{\alpha\bar{\beta}}$ vanishes for $1 \leq \alpha \leq s$ and $s+1 \leq \beta \leq r$ (with respect to local holomorphic frame e_α, $1 \leq \alpha \leq r$, with e_α in E' for $1 \leq \alpha \leq s$ and e_α, $1 \leq \alpha \leq r$, unitary at the point under consideration). The vanishing of such $\partial\ H_{\alpha\bar{\beta}}$ means that the orthogonal complement of E' in E is also a holomorphic subbundle and it would contradict the indecomposability of E into a nontrivial direct sum of holomorphic subbundles which are orthogonal to each other. Thus E is stable.

For the general case we first decompose E into nontrivial orthogonal direct summands (if possible) and the process of such a decompositon stops after a finite number of steps because of rank considerations. Then we apply the preceding argument to each indecomposable summand.

(1.7) *Remark.* We would like to observe that, for any given subbundle E' (or coherent subsheaf with torsion-free quotient) of E to have normalized first Chern number no more than that of E itself, it suffices to assume that the image of $(\Lambda F - \lambda I)|E'$ is orthogonal to E' instead of the stronger Hermitian-Einstein condition that $\Lambda F = \lambda I$ on E.

(1.8) For a Hermitian-Einstein vector bundle one has a Chern number inequality [Lü1]. First let us recall the definition of a Chern class. Suppose that X is an $r \times r$ matrix with elements in a commutative ring (*e.g.* the ring of even degree forms). Write

$$\det(I_r + X) = \Sigma_{k=0}^{r} \phi_k(X).$$

Then the k-th Chern class $c_k(E)$ of a Hermitian holomorphic vector bundle E of rank r and curvature form F is given by $\phi_k(\frac{i}{2\pi} F)$. In particular, $c_1(E) = \frac{i}{2\pi} \Sigma_\alpha F_{\alpha\bar{\alpha}}$ and

$$c_2(E) = (\tfrac{i}{2\pi})^2 \tfrac{1}{2} \Sigma_{\alpha,\beta}(F_{\alpha\bar{\alpha}} \wedge F_{\beta\bar{\beta}} - F_{\alpha\bar{\beta}} \wedge F_{\beta\bar{\alpha}}).$$

The Chern number inequality is that $((r-1)c_1(E)^2 - 2r\, c_2(E)) \wedge \omega^{m-2} \leq 0$.

To prove this we calculate $(r-1)c_1(E)^2 - 2r\, c_2(E)$ which is equal to

$$\begin{aligned} &(\tfrac{i}{2\pi})^2 \, (\Sigma_{\alpha,\beta}((r-1)F_{\alpha\bar{\alpha}} \wedge F_{\beta\bar{\beta}} + r(F_{\alpha\bar{\alpha}} \wedge F_{\beta\bar{\beta}} - F_{\alpha\bar{\beta}} \wedge F_{\beta\bar{\alpha}}))) \\ &= (\tfrac{i}{2\pi})^2 \, (-\Sigma_{\alpha,\beta}(F_{\alpha\bar{\alpha}} \wedge F_{\beta\bar{\beta}} + r\, F_{\alpha\bar{\beta}} \wedge F_{\beta\bar{\alpha}})) \\ &= (\tfrac{i}{2\pi})^2 (-(\mathrm{Tr}\, F)^2 + r\, \mathrm{Tr}(F^2)). \end{aligned}$$

To calculate $((r-1)c_1(E)^2 - 2r\, c_2(E)) \wedge \omega^{m-2}$, at the point under consideration we choose normal coordinates for the Kähler form ω so that $\omega = i\, \Sigma_{\alpha=1}^{m} dz^\alpha \wedge \overline{dz^\alpha}$. The coefficient of $(\frac{i}{2\pi})^2 \Pi_{\alpha=1}^{m} (i\, dz^\alpha \wedge \overline{dz^\alpha})$ in

$((r-1)c_1(E)^2 - 2r\, c_2(E))\wedge\omega^{m-2}$ is equal to

$$\Sigma_{\alpha,\beta,i,j}(-F_{\alpha\bar\alpha i\bar j}F_{\beta\bar\beta j\bar i} + F_{\alpha\bar\alpha i\bar i}F_{\beta\bar\beta j\bar j} + r\, F_{\alpha\bar\beta i\bar j}F_{\beta\bar\alpha j\bar i} - r\, F_{\alpha\bar\beta i\bar i}F_{\beta\bar\alpha j\bar j})$$
$$= \Sigma_{\alpha,\beta,i,j}(-F_{\alpha\bar\alpha i\bar j}F_{\beta\bar\beta j\bar i} + r\, F_{\alpha\bar\beta i\bar j}F_{\beta\bar\alpha j\bar i}) + (\Lambda\ \mathrm{Tr}\ F)^2 - r\ \mathrm{Tr}\ (\Lambda F)^2$$
$$= - \Sigma_{\alpha,\beta,i,j}\, F_{\alpha\bar\alpha i\bar j}F_{\beta\bar\beta j\bar i} + r\, \Sigma_{\alpha,i,j} F_{\alpha\bar\alpha i\bar j}F_{\alpha\bar\alpha j\bar i} + r\, \Sigma_{\alpha,\beta,i,j}|F_{\alpha\bar\beta i\bar j}|^2 + (\Lambda\ \mathrm{Tr}\ F)^2 - r\ \mathrm{Tr}\ (\Lambda F)^2$$
$$= \frac{1}{2}\Sigma_{\alpha,\beta,i,j}|F_{\alpha\bar\alpha i\bar j} - F_{\beta\bar\beta i\bar j}|^2 + r\, \Sigma_{\alpha,\beta,i,j}|F_{\alpha\bar\beta i\bar j}|^2 + (\Lambda\ \mathrm{Tr}\ F)^2 - r\ \mathrm{Tr}\ (\Lambda F)^2.$$

When we have a Hermitian-Einstein metric along the fibers of E, we have by definition $\Lambda F = \frac{1}{r}\mathrm{Tr}\ \Lambda F$. Hence in that case $(\Lambda\ \mathrm{Tr}\ F)^2 - r\ \mathrm{Tr}\ (\Lambda F)^2$ vanishes and $((r-1)c_1(E)^2 - 2r\, c_2(E))\wedge\omega^{m-2}$ is nonpositive.

For later use we would like to remark that for the general case we have

$$\text{(1.8.1)}\qquad \mathrm{Tr}(F^2)\wedge\omega^{m-2} = -\frac{(2\pi)^2}{r}\Big[\frac{1}{2}\Sigma_{\alpha,\beta,i,j}|F_{\alpha\bar\alpha i\bar j} - F_{\beta\bar\beta i\bar j}|^2 + r\,\Sigma_{\alpha,\beta,i,j}|F_{\alpha\bar\beta i\bar j}|^2 + (\Lambda\ \mathrm{Tr}\ F)^2 - r\ \mathrm{Tr}\ (\Lambda F)^2\Big]\wedge\omega^m + \frac{1}{r}(\mathrm{Tr}\ F)^2\wedge\omega^{m-2}$$
$$\le -\frac{(2\pi)^2}{r}\Big[(\Lambda\ \mathrm{Tr}\ F)^2 - r\ \mathrm{Tr}\ (\Lambda F)^2\Big]\wedge\omega^m + \frac{1}{r}(\mathrm{Tr}\ F)^2\wedge\omega^{m-2}.$$

(1.9) To construct a Hermitian-Einstein metric we start with an initial background Hermitian metric H_0 of E and then try to change H_0 to make it Hermitian-Einstein. So we need to know the relation between the curvuture tensor F_{H_0} of H_0 and the curvature tensor F_H of another Hermitian metric H.

Let $h = H H_0^{-1}$. In local coordinates $h_\alpha{}^\beta = H_{\alpha\bar\gamma}(H_0)^{\beta\bar\gamma}$ and h is a global section of the endomorphism bundle End(E) of E over M. Geometrically this means the following. From the relations

$$\langle u,v\rangle_H = H_{\alpha\bar\beta}\, u^\alpha\, \overline{v^\beta} = h_\alpha{}^\gamma (H_0)_{\gamma\bar\beta}\, u^\alpha\, \overline{v^\beta}$$
$$= (H_0)_{\gamma\bar\beta}\, (h_\alpha{}^\gamma u^\alpha)\, \overline{v^\beta} = \langle h\, u,v\rangle_{H_0}$$

and

$$\langle v,h\, u\rangle_{H_0} = \overline{\langle h\, u,v\rangle_{H_0}} = \overline{\langle u,v\rangle_H} = \langle v,u\rangle_H = \langle h\, v,u\rangle_{H_0}$$

we see that the endomorphism h of E is Hermitian with respect to the metric H_0 and this Hermitian endomorphism h expresses the metric H in terms of H_0. The endomorphism h is also Hermitian with respect to the metric H. The Hermitian property of h with respect to H_0 and H means that $h_\alpha{}^\gamma\, (H_0)_{\gamma\bar\beta}$ and $h_\alpha{}^\gamma\, H_{\gamma\bar\beta}$ both are Hermitian matrices.

We now use a local holomorphic basis of E. By definition we have

$$A_H - A_{H_0} = (\partial H)H^{-1} - (\partial H_0)H_0^{-1}$$
$$= (\partial h)H_0H_0^{-1}h^{-1} + (h\partial H_0)H_0^{-1}h^{-1} - (\partial H_0)H_0^{-1}$$
$$= \partial h\, h^{-1} + h(\partial H_0)H_0^{-1}h^{-1} - (\partial H_0)H_0^{-1}.$$

We use the notations ∂_H and $\bar\partial_H$ to denote respectively (1,0) and (0,1) exterior differentiation of E-valued or $\mathrm{End}(E)$-valued forms with respect to the complex metric connection A_H of H. In local coordinates the components of A_{H_0} are $(A_{H_0})_\alpha{}^\beta$. The differentiation of the endomorphism h is given in local coordinates by

$$(\partial_{H_0} h)_\alpha{}^\beta = \partial\, h_\alpha{}^\beta + h_\alpha{}^\gamma\, (A_{H_0})_\gamma{}^\beta - h_\gamma{}^\beta\, (A_{H_0})_\alpha{}^\gamma,$$

where the term $\partial\, h_\alpha{}^\beta$ means the (1,0) exterior differentiation of the *function* h_α^β which is the coefficient of the endomorphism h with respect to the local holomorphic basis of E. In matrix notations the equation reads

$$\partial_{H_0} h = \partial h + h\,A_{H_0} - A_{H_0} h$$
$$= \partial h + h(\partial H_0)H_0^{-1} - (\partial H_0)H_0^{-1} h.$$

It follows that

$$(\partial_{H_0} h)h^{-1} = A_H - A_{H_0}$$

and

(1.9.1) $$F_H = F_{H_0} + \bar{\partial}((\partial_{H_0} h)h^{-1}).$$

This relation between F_H and F_{H_0} holds for any pair of Hermitian metrics H and H_0 of E.

Later for the *a priori* estimate of h we need the following inequality

(1.9.2) $$\Delta \log \operatorname{Tr} h \geq -(|\Lambda F_{H_0}| + |\Lambda F_H|).$$

Now $\Delta \operatorname{Tr} h = -\operatorname{Tr} \Lambda \bar{\partial}\, \partial_{H_0} h$. From

$$F_H - F_{H_0} = \bar{\partial}((\partial_{H_0} h)h^{-1}) = (\bar{\partial}\partial_{H_0} h)h^{-1} + (\partial_{H_0} h)h^{-1}(\bar{\partial} h)h^{-1}$$

we have

$$(F_H - F_{H_0})h = \bar{\partial}\partial_{H_0} h + (\partial_{H_0} h)h^{-1}(\bar{\partial} h)$$

and $\Delta \operatorname{Tr} h = \operatorname{Tr} (\Lambda F_{H_0} - \Lambda F_H)h + \operatorname{Tr} \Lambda(\partial_{H_0} h)h^{-1}(\bar{\partial} h)$. Thus

$$\Delta \log \operatorname{Tr} h = \frac{\Delta \operatorname{Tr} h}{\operatorname{Tr} h} - \frac{|\nabla \operatorname{Tr} h|^2}{(\operatorname{Tr} h)^2}$$

$$\geq - (|\Lambda F_{H_0}| + |\Lambda F_H|) + (\operatorname{Tr} h)^{-1}\operatorname{Tr} \Lambda(\partial_{H_0} h)h^{-1}(\bar{\partial} h) - \frac{|\nabla \operatorname{Tr} h|^2}{(\operatorname{Tr} h)^2}.$$

We want to check that $(\operatorname{Tr} h)^{-1}\operatorname{Tr} \Lambda(\partial_{H_0} h)h^{-1}(\bar{\partial} h)$ dominates $\frac{|\nabla \operatorname{Tr} h|^2}{(\operatorname{Tr} h)^2}$. We choose local coordinates $z^1,\cdots,z^m$ normal at the point under consideration and also choose a local trivialization of E so that H_0 is the identity matrix at that point and dH_0 is zero at that point and H is diagnonal at that point. We get

$$\frac{|\nabla \operatorname{Tr} h|^2}{(\operatorname{Tr} h)^2} = \Sigma_{i=1}^m (\Sigma_{\alpha=1}^r h_\alpha{}^\alpha)^{-2} |\Sigma_{\alpha=1}^r \partial_i h_\alpha{}^\alpha|^2$$

$$= \Sigma_{i=1}^m (\Sigma_{\alpha=1}^r h_\alpha{}^\alpha)^{-2} |\Sigma_{\alpha=1}^r (\partial_i h_\alpha{}^\alpha)(h_\alpha{}^\alpha)^{-\frac{1}{2}} (h_\alpha{}^\alpha)^{\frac{1}{2}}|^2$$

$$\leq \Sigma_{i=1}^m (\Sigma_{\alpha=1}^r h_\alpha{}^\alpha)^{-2} \left[\Sigma_{\alpha=1}^r |\partial_i h_\alpha{}^\alpha|^2 (h_\alpha{}^\alpha)^{-1}\right]\left[\Sigma_{\alpha=1}^r h_\alpha{}^\alpha\right]$$

$$= \Sigma_{i=1}^m (\Sigma_{\alpha=1}^r h_\alpha{}^\alpha)^{-1} \left[\Sigma_{\alpha=1}^r |\partial_i h_\alpha{}^\alpha|^2 (h_\alpha{}^\alpha)^{-1}\right].$$

On the other hand,

$$(\operatorname{Tr} h)^{-1}\operatorname{Tr} \Lambda(\partial_{H_0} h)h^{-1}(\bar{\partial} h) = \Sigma_{i=1}^m (\Sigma_{\alpha=1}^r h_\alpha{}^\alpha)^{-1} \Sigma_{1\leq\alpha,\beta\leq r} (\partial_i h_\beta{}^\alpha)(h_\alpha{}^\alpha)^{-1}(\bar{\partial} h_\alpha{}^\beta)$$

$$\geq \Sigma_{i=1}^m (\Sigma_{\alpha=1}^r h_\alpha{}^\alpha)^{-1} \Sigma_{\alpha=1}^r (\partial_i h_\alpha{}^\alpha)(h_\alpha{}^\alpha)^{-1}(\bar{\partial} h_\alpha{}^\alpha)$$

$$\geq \frac{|\nabla \operatorname{Tr} h|^2}{(\operatorname{Tr} h)^2}.$$

This concludes the proof of (1.9.2). The inequality (1.9.1) is analogous to the estimation done by Aubin [A2, p.120, (β)] on $\Delta' \log(m+\Delta\varphi)$ in the proof of the existence of Kähler-Einstein metrics, because $\Delta' \log(m+\Delta\varphi)$ is the same as Δ' of the logarithm of the trace of the new Kähler metric with respect to the old one.

§2. *Gradient Flow and the Evolution Equation.*

(2.1) For the construction of Hermitian-Einstein metrics for stable bundles, we start with a fixed Hermitian metric H_0 and try to construct from it a Hermitian-Einstein metric by deforming H_0 through a one-parameter family of Hermitian metrics H_t $(0 \leq t < \infty)$. We sometimes suppress the subscript t in H_t and simply denote H_t by H. By using a conformal change, we can always assume without loss of generality that $\mathrm{Tr}\, F_H$ is harmonic with respect to the Kähler form ω for all values of t. To get a Hermitian-Einstein metric we want to make $\Lambda F - \lambda I$ vanish. To achieve this purpose we go along the gradient direction of the functional which is the global L^2 norm of $\Lambda F - \lambda I$ on M. Let us determine this gradient direction. First let us see how F_H changes when H changes as a function of the parameter t.

When we have a one-parameter family of metrics $H = H_t$, we can differentiate F_{H_t} in terms of t. From the identity (1.9.1) $F_{H_t} = F_{H_0} + \bar\partial((\partial_{H_0} h)h^{-1})$ it follows that $\frac{d}{dt} F_{H_t} = \bar\partial(\frac{\partial}{\partial t}((\partial_{H_0} h)h^{-1}))$. Now we evaluate $\frac{\partial}{\partial t}((\partial_{H_0} h)h^{-1})$. We have

$$\frac{\partial}{\partial t}((\partial_{H_0} h)h^{-1}) = \frac{\partial}{\partial t}\left[\partial h\, h^{-1} + h(\partial H_0)H_0^{-1}h^{-1} - (\partial H_0)H_0^{-1}\right]$$

$$= \frac{\partial}{\partial t}(\partial(hH_0)\, H_0^{-1}h^{-1}) = \partial((\frac{\partial h}{\partial t})H_0)H_0^{-1}h^{-1} + \partial(hH_0)H_0^{-1}(-h^{-1}\frac{\partial h}{\partial t}h^{-1})$$

$$= \partial(\frac{\partial h}{\partial t}h^{-1}) + \frac{\partial h}{\partial t}h^{-1}(\partial H)H^{-1} - (\partial H)H^{-1}\frac{\partial h}{\partial t}h^{-1} = \partial_H(\frac{\partial h}{\partial t}h^{-1}).$$

Hence $\frac{d}{dt} F_{H_t} = \bar\partial\partial_H((\frac{\partial h}{\partial t})h^{-1})$. To simplify notations we use $\dot h$ to denote $\frac{\partial h}{\partial t}$ and write

(2.1.1) $$\frac{d}{dt} F_{H_t} = \bar{\partial}\partial_H(\dot{h}h^{-1}).$$

(2.2) Let us now look at the gradient of the global L^2 norm of $\Lambda F - \lambda I$. We have

$$\begin{aligned}
\frac{d}{dt}(\Lambda F - \lambda I, \Lambda F - \lambda I) &= 2\,\mathrm{Re}(\Lambda \tfrac{\partial F}{\partial t}, \Lambda F - \lambda I) = 2\,\mathrm{Re}\,(\Lambda\bar{\partial}\partial_H(\dot{h}h^{-1}), \Lambda F - \lambda I) \\
&= 2\,\mathrm{Re}\int_M \nabla_{\bar{i}}\nabla_i(\dot{h}_\alpha{}^\gamma (h^{-1})_\gamma{}^\beta)\,\overline{(F_\alpha{}^\beta{}_{j\bar{j}} - \lambda\, I_\alpha{}^\beta)} \\
&= 2\,\mathrm{Re}\int_M (\dot{h}_\alpha{}^\gamma (h^{-1})_\gamma{}^\beta)\,\overline{\nabla_{\bar{i}}\nabla_i F_\alpha{}^\beta{}_{j\bar{j}}} \\
&= 2\,\mathrm{Re}\int_M (\dot{h}_\alpha{}^\gamma (h^{-1})_\gamma{}^\beta)\,\overline{\Delta(\Lambda F_\alpha{}^\beta)}.
\end{aligned}$$

If we would like to deform H in the opposite direction of the gradient of the global L^2 norm of $\Lambda F - \lambda I$, then we should choose the flow $\dot{h}\, h^{-1} = -\Delta(\Lambda F)$.

This flow is also the flow opposite to the gradient of the global L^2 norm of the full curvature tensor F. In this case we have

$$\begin{aligned}
\frac{d}{dt}(F,F) &= 2\,\mathrm{Re}(\tfrac{\partial F}{\partial t}, F) = 2\,\mathrm{Re}\,(\bar{\partial}\partial_H(\dot{h}h^{-1}), F) \\
&= 2\,\mathrm{Re}\int_M \nabla_{\bar{i}}\nabla_j(\dot{h}_\alpha{}^\gamma (h^{-1})_\gamma{}^\beta)\,\overline{F_\alpha{}^\beta{}_{j\bar{i}}} \\
&= 2\,\mathrm{Re}\int_M (\dot{h}_\alpha{}^\gamma (h^{-1})_\gamma{}^\beta)\,\overline{\nabla_{\bar{j}}\nabla_i F_\alpha{}^\beta{}_{j\bar{i}}} \\
&= 2\,\mathrm{Re}\int_M (\dot{h}_\alpha{}^\gamma (h^{-1})_\gamma{}^\beta)\,\overline{\nabla_{\bar{j}}\nabla_j F_\alpha{}^\beta{}_{i\bar{i}}} \quad \text{(by Bianchi's identity)} \\
&= 2\,\mathrm{Re}\int_M (\dot{h}_\alpha{}^\gamma (h^{-1})_\gamma{}^\beta)\,\overline{\Delta(\Lambda F_\alpha{}^\beta)}
\end{aligned}$$

and the equation for the flow opposite the direction of the gradient is still $\dot{h}\, h^{-1} = -\Delta(\Lambda F)$.

(2.3) The trouble with this flow is that the right-hand side is of order four. To do the analysis it is easier to deal with an equation of order two. Before we introduce the equation, let us look at the problem in a heuristic way and from a perspective that applies to a much more general setting. Our goal is to make $-(\Lambda F - \lambda I)$ vanish. Since we have already fixed a Hermitian metric H_0, instead of looking at the space of all Hermitian metrics H, we look at the space $\mathcal{H}$ of all endomorphisms $h = H H_0^{-1}$ of E which are positive definite with respect to H_0. For every $h \in \mathcal{H}$ we have an endomorphism $-(\Lambda F - \lambda I)h$ of E and we can regard $-(\Lambda F - \lambda I)h$ as a tangent vector of $\mathcal{H}$ at the point h. So we have a tangent vector field $-(\Lambda F - \lambda I)h$ on $\mathcal{H}$. We are looking for a point of $\mathcal{H}$ where this vector field vanishes. We can integrate the vector field and from any initial point get an integral curve given by $h = h(t)$, $0 \leq t < t_\infty$, where t_∞ is the maximum time-parameter value to which we can extend the integral curve. Suppose the maximum time-parameter value t_∞ is always ∞. Then we have two possiblities for $h(t)$ as $t \to \infty$. One is that the integral curve approaches the boundary of the space $\mathcal{H}$, that is, $h(t)$ becomes degenerate when $t \to \infty$. Another is that the velocity $(\Lambda F - \lambda I)h$ becomes zero as $h(t)$ approaches some point of $\mathcal{H}$ when $t \to \infty$. The second possibility is what we hope to get. Let us formulate this more precisely. The integral curve with the initial point $h(0) = I$ is given by $\frac{\partial}{\partial t} h(t) = -(\Lambda F_{H(t)} - \lambda I)h(t)$ with $h(0) = I$. Assuming that the maximum time-parameter value t_∞ is infinity means that the equation $\frac{\partial}{\partial t} h(t) = -(\Lambda F_{H(t)} - \lambda I)h(t)$ with $h(0) = I$ can be solved for all finite time t. Suppose $h(t)$ approaches some *nondegenerate* limit $h(\infty) \in \mathcal{H}$. We would like to see why $-(\Lambda F - \lambda I)h$ must vanish at $h(\infty)$. Since the equation $\frac{\partial}{\partial t} h(t) = -(\Lambda F_{H(t)} - \lambda I)h(t)$ does not depend on the independent variable t except through the dependent variable t, it follows that any translation $h(t + \alpha)$ of a solution $h(t)$ of the equation along the t-axis is also a solution of the equation. Since $h(t)$ approaches $h(\infty)$ as $t \to \infty$, it follows that the function $h(t + \alpha)$, $t_1 \leq t \leq t_2$, for any two finite values $t_1 < t_2$, approaches the *constant* function $h(\infty)$ defined on the interval $t_1 \leq t \leq t_2$ as $\alpha \to \infty$. The differential equation $\frac{\partial}{\partial t} h(t) = -(\Lambda F_{H(t)} - \lambda I)h(t)$ is equivalent to the integral equation $h(t) - h(t_1) = -\int_{\tau=t_1}^{t} (\Lambda F_{H(t)} - \lambda I)h(\tau)d\tau$. If the limit $h(\infty) = \lim_{t\to\infty} h(t)$ occurs in an appropriate space so that $F_{H(t)}$ approaches $F_{H(\infty)}$ as $t \to \infty$, then the *constant* function $h(\infty)$ defined on the interval $t_1 \leq t \leq t_2$ also satisfies the differential equation $\frac{\partial}{\partial t} h(t) = -(\Lambda F_{H(t)} - \lambda I)h(t)$ It now follows from the constancy of the function $h(\infty)$ that $\Lambda F_{H(\infty)} - \lambda I = 0$.

The above method of finding a zero of a vector field by using an evolution equation going along the flow applies to very general situations. One can either go along the flow or opposite to it. The evolution equation is some parabolic equation like the heat equation and there is only one direction for which we have a solution for all finite time and we must go along that direction. In our case the differential operator ΛF on H is like the positive Laplacian. So we know that we should consider the equation $\frac{\partial h}{\partial t} = -(\Lambda F - \lambda I)h$ instead of the equation $\frac{\partial h}{\partial t} = (\Lambda F - \lambda I)h$.

At this point we should explain why we use the vector field $-(\Lambda F - \lambda I)h$ instead of simply the vector field $-(\Lambda F - \lambda I)$ since our goal is to determine the point h where $-(\Lambda F - \lambda I)$ vanishes. The reason is that we want to preserve the condition that the Chern form of the determinant line bundle $\det E$ of E for the metric $\det H$ is harmonic. Since $\frac{\partial}{\partial t} \log \det h = \operatorname{Tr}(\frac{\partial h}{\partial t} h^{-1})$ and $-(\Lambda F - \lambda I)$ is trace-free, we can guarantee $\det h \equiv 1$ by using the equation $\frac{\partial h}{\partial t} h^{-1} = -(\Lambda F - \lambda I)$. Another reason is that since $\Lambda F = -g^{i\bar{j}}\partial_i((\partial_{\bar{j}}H)H^{-1})$, to make our evolution equation of the same type as the heat equation the right-hand side should be $\frac{\partial h}{\partial t} h^{-1}$ instead of just $\frac{\partial h}{\partial t}$.

§3 *Existence of Solution of Evolution for Finite Time.*

(3.1) Now we want to show the existence of solution of the evolution equation $\frac{\partial h}{\partial t} h^{-1} = -(\Lambda F - \lambda I)$ for all finite time. We require that det h *is always the constant function 1*. This sort of normalization is necessary, because the equation is unchanged by multiplying h by a positive constant. We use the continuity method. Openness follows from the solvability of a parabolic equation for *small* time. For closedness we need *a priori* estimates of F as t approaches some finite limit T. These estimates are obtained from the maximum principle of the heat equation. For this purpose we show that the norms of F and other tensors associated with F satisfy inequalities involving the heat operator. We have to first calculate the time-derivative of F and the Laplacian of F. We use the notation $\dot{F}$ to denote the time-derivative $\frac{\partial F}{\partial t}$ of F. We have

$$\dot{F} = \bar{\partial}\partial_H(\dot{h}h^{-1}) = -\bar{\partial}\partial_H(\Lambda F - \lambda I) = -\bar{\partial}\partial_H(\Lambda F).$$

Hence we have $\Lambda\dot{F} = -\Lambda\bar{\partial}\partial_H(\Lambda F) = \Delta(\Lambda F)$ and $(\frac{\partial}{\partial t} - \Delta)\Lambda F = 0$.

One consequence of the above computation of $\dot{F}$ which is not needed right away but will be needed later is that the global L^2 norm of the full curvature tensor F is nonincreasing as a function of t, because

$$\begin{aligned}\frac{d}{dt}(F,F) &= 2\,\mathrm{Re}(\dot{F},F) = 2\,\mathrm{Re}(\bar{\partial}\partial_H(\Lambda F),F)\\ &= 2\,\mathrm{Re}\int_M \nabla_{\bar{k}}\nabla_{\ell}(F_{\alpha}{}^{\beta}{}_{i\bar{i}})\cdot F_{\beta}{}^{\alpha}{}_{k\bar{\ell}}\\ &= 2\,\mathrm{Re}\int_M \nabla_{\bar{k}}\nabla_{i}(F_{\alpha}{}^{\beta}{}_{\ell\bar{i}})\cdot F_{\beta}{}^{\alpha}{}_{k\bar{\ell}}\\ &= -\,2\,\mathrm{Re}\int_M \nabla_{i}(F_{\alpha}{}^{\beta}{}_{\ell\bar{i}})\cdot \nabla_{\bar{k}}F_{\beta}{}^{\alpha}{}_{k\bar{\ell}} \le 0.\end{aligned}$$

(3.2) We now continue with our estimates of the norms of the curvature tensor. Let $e = |F|^2$, $\hat{e} = |\Lambda F|^2$, and $e_k = |\nabla_H^k F|^2$ for $k \ge 0$. From $(\frac{\partial}{\partial t} - \Delta)\Lambda F = 0$ we have

$$(\frac{\partial}{\partial t} - \Delta)\hat{e} = 2\,\mathrm{Re}((\frac{\partial}{\partial t} - \Delta)\Lambda F, \Lambda F) - \|\nabla(\Lambda F)\|^2 \leq 0,$$

$$\frac{\partial}{\partial t}|F|^2 = 2\,\mathrm{Re}(\frac{\partial F}{\partial t}, F) = 2\,\mathrm{Re}(\bar{\partial}\partial_H(\Lambda F), F),$$

$$(\bar{\partial}\partial_H(\Lambda F))_{i\bar{j}} = \nabla_{\bar{i}}\nabla_j F_{k\bar{k}} = \nabla_{\bar{i}}\nabla_k F_{j\bar{k}} = \nabla_k\nabla_{\bar{i}}F_{j\bar{k}} + \{F,F\} + \{R,F\},$$

$$(\nabla_k\nabla_{\bar{i}}F_{j\bar{k}}, F) = \int_M \nabla_k\nabla_{\bar{i}}F_{j\bar{k}}\,\overline{F_{i\bar{j}}} = \int_M \nabla_k\nabla_{\bar{k}}F_{j\bar{i}}\,\overline{F_{i\bar{j}}}$$
$$= -\int_M \nabla_{\bar{k}}F_{j\bar{i}}\,\overline{\nabla_{\bar{k}}F_{i\bar{j}}}.$$

By using the Bianchi identity and the commutation formulae, we get

$$(\frac{\partial}{\partial t} - \Delta)e \leq \{R,F,F\} + \{F,F,F\} \leq C(e^{3/2} + e).$$

Here $\{R,F\}$ means some expression linear in the curvauture tensor R of M and the curvature tensor F of E and the coefficients in the expression are some universal constants. The expressions $\{F,F\}$, $\{R,F,F\}$, $\{F,F,F\}$ carry similar meanings. They all come from the commutation formulas for covariant differentiation. By commuting derivatives, we find that

$$(\frac{\partial}{\partial t} - \Delta)e_k \leq c_k e_k^{1/2}(\Sigma_{i+j=k}\, e_i^{1/2}(e_j^{1/2} + 1)).$$

(3.3) Because of the inequality $(\frac{\partial}{\partial t} - \Delta)\hat{e} \leq 0$ we conclude from the maximum principle for the heat operator that $\sup_M \hat{e}$ is bounded uniformly in t for $t < \infty$.

It is more difficult to use the equation

$$(\frac{\partial}{\partial t} - \Delta)e_k \leq c_k e_k^{1/2}(\Sigma_{i+j=k}\, e_i^{1/2}(e_j^{1/2} + 1))$$

because the right-hand side involves a $\frac{3}{2}$ power. For an estimate the most one can allow is a first power, otherwise we do not have linearity anymore. For example, one can consider the heat equation

$$\frac{\partial f}{\partial t} - \Delta f = C(1 + f), \quad f(0) = e_k(0).$$

This linear heat equation has smooth solution f defined for all $t \geq 0$. If we have an inequality

$$\left(\frac{\partial}{\partial t} - \Delta\right)e_k \leq C(1 + e_k),$$

then by applying the maximum principle to the function $(e_k - f)e^{-Ct}$ one gets a bound for e_k. To get rid of this $\frac{3}{2}$ power, *we assume that the curvature* F *is uniformly bounded for* $t < T$. We claim that under this assumption the derivative of any fixed order of F is uniformly bounded for $t < T$. The reason is that under this assumption we conclude by induction on k that the inequality

$$\left(\frac{\partial}{\partial t} - \Delta\right)e_k \leq C(1 + e_k)$$

holds.

(3.4) So to get the existence of the solution for the heat equation for any *finite* time we have to worry about the uniform bound for the full curvature tensor F for $t < T$. First we are going to use the inequality $\left(\frac{\partial}{\partial t} - \Delta\right)e \leq C(e^{3/2} + e)$ to reduce the requirement of uniform bound to that of some L^p bound for $t < T$. We do this by using the heat kernel. The heat kernel $H_t(x,y)$ for the heat operator $\frac{\partial}{\partial t} - \Delta$ on M is given for small times t and nearby points $x,y \in M$ by $ct^{-m}e^{-r^2/4t}$ with $r = \mathrm{dist}(x,y)$ up to the addition of a smooth function. Letting $u = \frac{r}{2\sqrt{pt}}$, we have

$$(t^{-m}e^{-r^2/4t})^p\ r^{2m-1}dr = c\ t^{-mp}e^{-u^2}\ t^{(2m-1)/2}\ r^{2m-1}\ t^{1/2}\ du$$
$$= c'\, t^{m(1-p)}e^{-u^2}du$$

and

$$\|H_t(x,\cdot)\|_{L^p(M)} \le Ct^{m(1-p)/p}$$

which is $\le C't^{-\mu}$ for some $\mu < 1$ near $t = 0$ if $p < \frac{m}{m-1}$. So when we have $p < \frac{m}{m-1}$,

$$\int_{t=0}^{T} \|H_t(x,\cdot)\|_{L^p(M)}\, dt \le c_p(T).$$

From $(\frac{\partial}{\partial t} + \Delta)e \le C(e^{3/2} + e)$ we have

$$e_t \le H_t \cdot e_0 + c \int_0^t H_{t-\tau} \cdot (e_\tau^{3/2} + e_\tau)\, d\tau.$$

The first term is bounded by $\sup_M e_0$ and the second one is bounded if the $L^q(M)$ norm of $e^{3/2}$ is bounded when $p < \frac{m}{m-1}$, where $\frac{1}{p} + \frac{1}{q} = 1$. This means that if the $L^r(M)$ norm of F is bounded for $r > 3m$, then we have the supremum norm bound for F.

(3.5) We can now finish the proof of the existence of the solution for the heat equation for finite time. Suppose H_t, $0 \le t < T$, is a one-parameter family of metrics along the fibers of a holomorphic bundle E over M such that (i) H_t converges in C^0 norm to some continuous metric H_T as $t \to T$; (ii) the supremum norm over M of ΛF is bounded uniformly for $t < T$. We claim that because of the elliptic equation expressing the contracted curvature tensor ΛF in terms of H_t, the metrics H_t are actually bounded in C^1 norm and the *full* curvature tensor F_{H_t} is bounded in L^p norm for any finite p uniformly in $t < T$.

We verify this by the argument of absurdity. Suppose that the metrics H_t are not bounded uniformly in C^1 norm so that for some sequence $t_i \to T$ there are points $x_i \in M$ such that the supremum norm m_i of ∇H_i is achieved at x_i and $m_i \to \infty$, where we have used the simpler notation H_i to denote H_{t_i}. By taking a subsequence of x_i we can assume without loss of generality that x_i converges to some point in M. Let D_r denote the polydisk consisting of all $z = (z^1, \cdots, z^m) \in \mathbb{C}^m$ such that $|z^\alpha| < r$ for $1 \leq \alpha \leq m$. Since this is a purely local problem, we can choose local coordinates z^α in the polydisk $D_1 = \{|z^\alpha| < 1\}$ and regard H_i as a matrix-valued function in z^α. After a slight translation of the coordinates we can assume that $\sup_{D_1} |\nabla H_i| = m_i$ is achieved at $z = 0$ for all i. Let $\tilde{H}_i(z) = H_i(\frac{z}{m_i})$. Then $\sup_{D_1} |\nabla \tilde{H}_i| = 1$ is achieved at $z = 0$. Since

$$\Lambda F_{H_i} = (\Delta H_i)H_i^{-1} - i\Lambda\partial H_i H_i^{-1}\bar{\partial} H_i H_i^{-1}$$

and $\Delta\tilde{H}_i(z) = m_i^{-2}(\Delta H_i)(z/m_i)$ and $\partial\tilde{H}_i(z) = m_i^{-2}(\partial H_i)(z/m_i)$ and $\bar{\partial}\tilde{H}_i(z) = m_i^{-2}(\bar{\partial} H_i)(z/m_i)$, it follows that

$$\left[(\Delta\tilde{H}_i)\tilde{H}_i^{-1} - i\Lambda\partial\tilde{H}_i\tilde{H}_i^{-1}\bar{\partial}\tilde{H}_i\tilde{H}_i^{-1}\right](z) = m_i^{-2}(\Lambda F_{H_i})(z/m_i)$$

is uniformly bounded in D_1, because $(\Lambda F_{H_i})(z)$ is uniformly bounded on D_1. Since both $\tilde{H}_i$ and $\nabla\tilde{H}_i$ are uniformly bounded on D_1, it follows that $\Delta\tilde{H}_i$ is uniformly bounded on D_1. By elliptic estimates we know that $\tilde{H}_i$ is bounded in L_2^p norm on $D_{1/2}$ for any $p < \infty$. (The shrinking D_1 to $D_{1/2}$ is used to make sure that it is an interior elliptic estimate.) For $p > 2m$ the inclusion $L_2^p \subset C^1$ is a compact operator and we can find a subsequence of i so that $\tilde{H}_i$ converges in C^1 to some $\tilde{H}_\infty$ on $D_{1/3}$. On the other hand that from the C^0 converges of H_i to some H_∞ that $\tilde{H}_i$ converges to the constant matrix $H_\infty(0)$ in C^0 norm on $D_{1/3}$. Thus $\tilde{H}_\infty$ equals the constant matrix $H_\infty(0)$. Thus $\nabla\tilde{H}_i(0)$ converges to $\nabla\tilde{H}_\infty(0) = 0$, contradicting the fact that $|\nabla\tilde{H}_i(0)| = 1$ for all i. Hence the metrics H_t are bounded uniformly in C^1 norm. From

$$\Lambda F_{H_i} = (\Delta H_i)H_i^{-1} - i\Lambda\partial H_i H_i^{-1}\bar{\partial} H_i H_i^{-1}$$

we conclude that ΔH_i is bounded in C^0 norm and by the ellitpic estimates for Δ we conclude that H_i is bounded in L_2^p norm for any $p < \infty$. Hence F_{H_i} is bounded in L^p norm for any $p < \infty$.

This argument of obtaining the L_2^p bound and the C^1 bound of H_t and the L^p bound of F_{H_t} form the uniform bound of H_t and the uniform bound of ΛF_{H_t} works also in the case of $T = \infty$. For later application in the case $T = \infty$ we would like to change the assumption of (i) and replace it by the following condition: (i)' Both the C^0 norm of H_t and the L_1^2 norm of H_t are bounded in independent of t and there is a positive lower bound for the eigenvalues of H_t independent of t. For this change of assumption we need only verify that under the new assumption (i)', for any sequence H_{t_i} with $t_i \to \infty$ we can select a subsequence t_{i_ν} such that $H_{t_{i_\nu}}$ converges in C^0 to some continuours Hermitian metric H_∞ of E. Since the L_1^2 norm of H_t is bounded independent of t, we can choose a subsequence of t_i (which we denote again by t_i wihtout loss of generality) that H_{t_i} converges strongly in L^2 norm. By (1.9.2) we have

$$\Delta \log \mathrm{Tr}((H_{t_i})H_{t_j}^{-1}) \geq -(|\Lambda F_{H_{t_i}}| + |\Lambda F_{H_{t_j}}|)$$

from which it follows that

$$\Delta \mathrm{Tr}((H_{t_i})H_{t_j}^{-1}) \geq -(|\Lambda F_{H_{t_i}}| + |\Lambda F_{H_{t_j}}|)\ \mathrm{Tr}((H_{t_i})H_{t_j}^{-1}) \geq -C,$$

where C is a constant independent of i and j. Thus

$$\Delta \left[\mathrm{Tr}((H_{t_i})H_{t_j}^{-1}) + \mathrm{Tr}((H_{t_j})H_{t_i}^{-1}) - 2\right] \geq -2C.$$

Since H_{t_i} converges strongly in L^2 as $i \to \infty$, it follows that $\mathrm{Tr}((H_{t_i})H_{t_j}^{-1}) + \mathrm{Tr}((H_{t_j})H_{t_i}^{-1}) - 2$ converges strongly in L^1 to 0 as both i and j go to infinity. Let $G(P,Q)$ be the Green's function for M so that

$$\varphi(P) = \frac{1}{\mathrm{Vol}\ M}\int_{Q\in M}\varphi(Q)dQ + \int_{Q\in M} G(P,Q)(-\Delta\varphi)(Q)dQ$$

for any smooth function φ on M. Let $-K$ be a negative lower bound of $G(P,Q)$ for P, $Q \in M$. Then

$$\begin{aligned}\varphi(P) &= \frac{1}{\mathrm{Vol}\ M}\int_{Q\in M}\varphi(Q)dQ + \int_{Q\in M}(G(P,Q) + K)(-\Delta\varphi)(Q)dQ \\ &\leq \frac{1}{\mathrm{Vol}\ M}\int_{Q\in M}\varphi(Q)dQ + \sup_{Q\in M}(-\Delta\varphi)\int_{Q\in M}(G(P,Q) + K)dQ \\ &= \frac{1}{\mathrm{Vol}\ M}\int_{Q\in M}\varphi(Q)dQ + \sup_{Q\in M}(-\Delta\varphi)\ K\ (\mathrm{Vol}\ M).\end{aligned}$$

We apply this to the case $\varphi = \mathrm{Tr}((H_{t_i})H_{t_j}^{-1}) + \mathrm{Tr}((H_{t_j})H_{t_i}^{-1}) - 2$ and conclude that the C^0 norm of $\mathrm{Tr}((H_{t_i})H_{t_j}^{-1}) + \mathrm{Tr}((H_{t_j})H_{t_i}^{-1}) - 2$ approaches to zero as i and j go to infinity. Thus H_{t_i} approaches in C^0 norm some continuous Hermitian metric H_∞ of E as $i \to \infty$.

(3.6) For the existence of the heat equation for finite time the only thing left to prove is the C^0 converges of H_t as t approaches T from the left through a suitable subsequence. For if we have this then we know that F_{H_t} is bounded in C^k norm for any k and from

$$\Lambda F_{H_t} = (\Delta H_t)H_t^{-1} - i\Lambda\partial H_t H_t^{-1}\bar{\partial} H_t H_t^{-1}$$

and the elliptic estimate of Δ that H_t is bounded in C^k norm for any k.

(3.7) For two Hermitian metrics H, K let $\tau(H,K) = \mathrm{Tr}(HK^{-1})$ and $\sigma(H,K) = \tau(H,K) + \tau(K,H) - 2\,\mathrm{rank}\,E$. We claim that if H and K are two solutions of the evolution equation, then $\frac{\partial\sigma}{\partial t} - \Delta\sigma \leq 0$. It suffices to check $\frac{\partial\tau}{\partial t} - \Delta\tau \leq 0$. Now

$$\begin{aligned}\frac{\partial\tau}{\partial t} &= \mathrm{Tr}\left(\frac{\partial H}{\partial t}K^{-1} - HK^{-1}\frac{\partial K}{\partial t}K^{-1}\right)\\ &= \mathrm{Tr}(-(\Lambda F_H - \lambda I)HK^{-1} + HK^{-1}(\Lambda F_K - \lambda I))\\ &= \mathrm{Tr}((\Lambda F_K - \Lambda F_H)\, h),\end{aligned}$$

where $h = HK^{-1}$. Since $F_H = F_K + \bar{\partial}_K((\partial_K h)h^{-1})$, it follows that

$$\begin{aligned}\frac{\partial\tau}{\partial t} &= -\mathrm{Tr}(\Lambda\bar{\partial}_K((\partial_K h)h^{-1})h)\\ &= \mathrm{Tr}(\Delta_K h - \Lambda(\partial_K h)h^{-1}(\bar{\partial}_K h))\\ &= \Delta\tau - \mathrm{Tr}(\Lambda(\partial_K h)h^{-1}(\bar{\partial}_K h))\end{aligned}$$

and $\frac{\partial\tau}{\partial t} - \Delta\tau \leq 0$.

When H_t is a solution of the evolution equation, $H_{t-\delta}$ is also a solution of the evolution equation. Apply the above result to $K_t = H_{t-\delta}$. Then by the maximum principle for the heat equation we have $\sup_M \sigma(H_t, H_{t-\delta}) \leq \sup_M \sigma(H_{t_0}, H_{t_0-\delta})$ for $t \leq t_0$. Fix t_0 and for any given $\epsilon > 0$ choose $\delta > 0$ such that $\sup_M \sigma(H_{t_0}, H_s) < \epsilon$ for $|s - t_0| < \delta$. Then $\sup_M \sigma(H_t, H_s) < \epsilon$ for s, t in $(T-\delta, T)$. Hence H_t is uniformly Cauchy as t approaches T from the left. Thus we have the C^0 convergence of H_t and we have a proof of the existence of the solution of the evolution equation for any finite time.

§4. *Secondary Characteristics.*

(4.1) To get the convergence of the solution of the heat equation at infinite time, we have to use the assumption of stability of the bundle. The use of the assumption of the stability of the bundle is done by induction on the dimension of the base manifold by using a functional on the space of Hermitian metrics which we dub the *Donaldson functional*. This functional is essentially the potential function of the vector field $(\Lambda F - \lambda I)h$. To obtain a manageable explicit form of this functional, we have to introduce secondary characteristic classes.

Recall that heuristically our evolution equation is obtained by integrating a vector field on the space of Hermtian metrics and the fixed points of this vector field are the Hermitian-Einstein metrics. In our heuristic discussion the vector field was not presented as the gradient vector field of some functional. As a matter of fact, as we discussed before, if we consider the functional of the square norm of the full curvature tensor F or of $\Lambda F - \lambda I$, we would get a similar vector field whose evolution equation is of order four.

Heuristically if we want to reduce the order of the evolution equation by two, we should use a functional that involves not some expression of the curvature tensor but some expression of entities whose second-order derivatives are expressions of the curvature tensor. The expressions of the curvature tensor we use are the Chern forms and the entities whose second derivatives are the Chern forms are the *secondary characteristic classes* which we are now going to introduce.

(4.2) Given two Hermitian metrics H, H' along the fibers of E, we have two Chern forms of type (p,p) (given by the elementary symmetric functions of the "eigenvalues" of the curvature forms). These two Chern forms are in the same (p,p) class and differ only by $\partial\bar{\partial}$ of a (p-1,p-1)-form. This (p-1,p-1)-form is called the secondary characteristic. Let us look at the the case of the first and second Chern forms. Since

$$c_1(E) = \mathrm{Tr}\, F = \bar{\partial}\partial \log \det H,$$

we have

$$\mathrm{Tr}\, F_H - \mathrm{Tr}\, F_{H_0} = \bar{\partial}\partial \log (\det H / \det H_0) = \bar{\partial}\partial \log \det(H\, H_0^{-1}).$$

The secondary Chern class is $-\log \det(H\, H_0^{-1})$. We define $R_1 = \log \det(H\, H_0^{-1})$. If H depends on a real parameter t, then $\frac{\partial}{\partial t} R_1 = \mathrm{Tr}(\dot{h}\, h^{-1})$, where $h = H\, H_0^{-1}$ and the overhead dot means differentiation with respect to t. When we restrict ourselves to Hermitian metrics H with the property $\det h = 1$, the function R_1 is simply identically zero.

(4.3) We use the notations of (2.1). Since $\frac{\partial}{\partial t} F = \bar{\partial}\partial_H(\dot{h}\, h^{-1})$ by (2.1.3), it follows that $\frac{\partial}{\partial t}\mathrm{Tr}(F\wedge F) = 2\, \mathrm{Tr}(F\wedge\bar{\partial}\partial_H(\dot{h}h^{-1})) = 2\, \bar{\partial}\partial\, \mathrm{Tr}(F\dot{h}h^{-1})$, because the vanishing of $\partial_H F$ and $\bar{\partial}F$ come from the Bianchi identity. So the secondary characteristic for the second Chern class is given by an expression involving $\int_0^t \mathrm{tr}(F\dot{h}h^{-1})dt$. We define R_2 to be $\sqrt{-1}\int_0^t \mathrm{tr}(F\dot{h}h^{-1})dt$ so that

$$\frac{\partial}{\partial t} R_2 = \sqrt{-1}\ \mathrm{Tr}(F\dot{h}h^{-1}).$$

§5. *Donaldson's Functional.*

(5.1) We are now ready to define Donaldson's functional. It is given by

$$\mathcal{M} = \int_M m\ R_2 \Lambda\omega^{m-1} - \lambda R_1 \omega^m.$$

It is defined so that

$$\frac{d\mathcal{M}}{dt} = \int_M m\sqrt{-1}\ \mathrm{Tr}(F\dot{h}h^{-1})\wedge\omega^{m-1} - \lambda(\mathrm{Tr}\ \dot{h}h^{-1})\omega^m$$
$$= \int_M \mathrm{Tr}((\Lambda F - \lambda I)\dot{h}h^{-1})\omega^m.$$

This gives the gradient of $\mathcal{M}$ on the space of Hermitian metrics. The curve of steepest descent for $\mathcal{M}$ is $\dot{h}h^{-1} = -(\Lambda F - \lambda I)$. The Donaldson functional $\mathcal{M}$ takes the place of the global square norm of the full curvature tensor F or of $\Lambda F - \lambda I$. The curve of the steepest descent of the global square norm of the full curvature tensor F or of $\Lambda F - \lambda I$ is given by $\dot{h}h^{-1} = -\Delta(\Lambda F)$ which is a fourth-order parabolic equation. Now by using the Donaldson functional we have a parabolic equation of order two $\dot{h}h^{-1} = -(\Lambda F - \lambda I)$ for the curve of steepest descent. This is the reason for the introduction of the Donaldson functional.

The definition of $\mathcal{M}$ involves the secondary characteristics R_1 and R_2. The secondary characteristic R_1 is a function of H and H_0. However, R_2 is defined by integrating along a path joining H_0 to H. We claim that $\mathcal{M}$ is independent of the choice of path when we integrate from H_0 to H. This means that $\mathcal{M}$ is the potential function of some conservative vector field on the space of Hermitian metrics.

For notational simplicity we use subscripts s and t to denote differentiation with respect to s and t respectively.

$$\frac{\partial}{\partial s}\operatorname{Tr}(Fh_t h^{-1}) = \operatorname{Tr}(F_s h_t h^{-1} + Fh_{st}h^{-1} - Fh_t h^{-1}h_s h^{-1})$$
$$= \operatorname{Tr}(\bar{\partial}\partial_H(h_s h^{-1})h_t h^{-1} + Fh_{st}h^{-1} - Fh_t h^{-1}h_s h^{-1})$$

$$\frac{\partial}{\partial s}\operatorname{Tr}(Fh_t h^{-1}) - \frac{\partial}{\partial t}\operatorname{Tr}(Fh_s h^{-1})$$
$$= \operatorname{Tr}(\bar{\partial}\partial_H(h_s h^{-1})h_t h^{-1} - Fh_t h^{-1}h_s h^{-1} - \bar{\partial}\partial_H(h_t h^{-1})h_s h^{-1} + Fh_s h^{-1}h_t h^{-1}).$$

Now use the commutation formula

$$\bar{\partial}\partial_H(h_s h^{-1}) = -\partial_H\bar{\partial}(h_s h^{-1}) - Fh_s h^{-1} + h_s h^{-1}F$$

to get

$$\frac{\partial}{\partial s}\operatorname{Tr}(Fh_t h^{-1}) - \frac{\partial}{\partial t}\operatorname{Tr}(Fh_s h^{-1})$$
$$= \operatorname{Tr}(-\partial_H\bar{\partial}(h_s h^{-1})h_t h^{-1} + h_s h^{-1}Fh_t h^{-1} - Fh_t h^{-1}h_s h^{-1} - \bar{\partial}\partial_H(h_t h^{-1})h_s h^{-1})$$
$$= \operatorname{Tr}(-\partial_H\bar{\partial}(h_s h^{-1})h_t h^{-1} - \bar{\partial}\partial_H(h_t h^{-1})h_s h^{-1})$$

because $\operatorname{Tr}(h_s h^{-1}Fh_t h^{-1}) = \operatorname{Tr}(Fh_t h^{-1}h_s h^{-1})$. Thus by integration by parts we have

$$\int_M \frac{\partial}{\partial s}\operatorname{Tr}(Fh_t h^{-1})\wedge\omega^{m-1} - \int_M \frac{\partial}{\partial t}\operatorname{Tr}(Fh_s h^{-1})\wedge\omega^{m-1}$$
$$= \int_M \operatorname{Tr}(-\partial_H\bar{\partial}(h_s h^{-1})h_t h^{-1} - \bar{\partial}\partial_H(h_t h^{-1})h_s h^{-1})\wedge\omega^{m-1}$$
$$= \int_M \operatorname{Tr}(\bar{\partial}(h_s h^{-1})\partial_H h_t h^{-1} + \partial_H(h_t h^{-1})\bar{\partial}h_s h^{-1})\wedge\omega^{m-1} = 0$$

because $\operatorname{Tr}(A\wedge B + B\wedge A) = 0$ when A and B are matrices of *1-forms*. If we have two paths joining H_0 to H, then we have $h(s,t)$ with $h(s,0) = I$ and $h(s,1) = HH_0^{-1}$ so that $h(0,t)$ and $h(1,t)$ are the two paths joining I to HH_0^{-1} and

$$\frac{d}{ds}\int_M R_2\wedge\omega^{m-1} = \sqrt{-1}\,\frac{d}{ds}\int_{t=0}^{1}\int_M \operatorname{Tr}(Fh_t h^{-1})\wedge\omega^{m-1}$$
$$= \sqrt{-1}\int_{t=0}^{1}\int_M \frac{\partial}{\partial s}\operatorname{Tr}(Fh_t h^{-1})\wedge\omega^{m-1}$$
$$= \sqrt{-1}\int_{t=0}^{1}\int_M \frac{\partial}{\partial t}\operatorname{Tr}(Fh_s h^{-1})\wedge\omega^{m-1}$$
$$= \sqrt{-1}\int_M \operatorname{Tr}(Fh_s h^{-1})\wedge\omega^{m-1}\Big|_{t=0}^{1} = 0,$$

because h_s vanishes both at $t = 0$ and $t = 1$.

(5.3) The Donaldson functional will be used to take advantage of the assumption of the stability of the bundle in the induction process on the dimension of the base manifold. So we want to see how the Donaldson functional for M is related to the Donaldson functional for a hypersurface M' of M. The property we will need is that the Donaldson functional for a hypersurface is estimated from above by that of the Donaldson functional for the ambient manifold. This property is proved by an argument similar to that for the adjunction formula.

Suppose the Kähler form ω is the curvature form of some Hermitian line bundle L. We take a hypersurface M' of M in the class $\mu\,\omega$. On M' we have the Donaldson functional $\mathcal{M}_{M'} = \int_{M'} (m-1)R_2\wedge\omega^{m-2} - \lambda R_1\omega^{m-1}$ for the algebraic manifold M'. We assume that $\det H = \det H_0$ and that the uniform norm of ΛF is bounded. We would like to show that the Donaldson functional $\mathcal{M}_{M'}$ for M' is bounded from above by a constant multiple of the Donaldson functional $\mathcal{M}_M$ for M plus a constant.

Now M' is defined by a holomorphic section s of the line bundle $L^{\otimes\mu}|M'$. We assume (after multiplying it by a small positive constant) that the pointwise length of s is less than 1 everywhere on M. Let $f = \frac{1}{2\pi\mu}\log|s|^2$. As a current f is negative and $\sqrt{-1}\partial\bar\partial f = \frac{1}{\mu}[M'] - \frac{1}{2\pi}\,\omega$

on M. Hence

$$\frac{1}{2\pi}\int_M (m\ R_2 - \lambda R_1\omega)\omega^{m-1} = \int_M (m\ R_2 - \lambda R_1\omega)(\frac{1}{\mu}[M'] - \sqrt{-1}\partial\bar{\partial}f)\omega^{m-2}$$

$$= \frac{m}{\mu(m-1)}\int_{M'} ((m-1)R_2 - \lambda R_1\omega)\omega^{m-2} + \frac{\lambda}{\mu(m-1)}\int_M R_1\omega^m - \int_M (m\ R_2 - \lambda R_1\omega)(\sqrt{-1}\partial\bar{\partial}f)\omega^{m-2}.$$

So

$$\frac{1}{2\pi}\mathcal{M}_M = \frac{m}{\mu(m-1)}\mathcal{M}_{M'} + \frac{\lambda}{\mu(m-1)}\int_M R_1\omega^m - \int_M f(\sqrt{-1}\partial\bar{\partial}(m\ R_2 - \lambda R_1\omega))\omega^{m-2}$$

$$= \frac{m}{\mu(m-1)}\mathcal{M}_{M'} + \frac{\lambda}{\mu(m-1)}\int_M R_1\omega^m - \int_M f(\psi(F_H) - \psi(F_{H_0}))\omega^{m-1},$$

where $\psi(F_H) = \frac{m}{2}Tr(F_H^2) + \lambda\sqrt{-1}Tr(F_H)\omega$ from the definition of R_2 and R_1.

Since $\det H = \det H_0$, we have $Tr(F_H) = Tr(F_{H_0})$ and $\int_M f\ Tr(F_H)\ \omega^m$ is bounded by a constant. We now use the inequality (1.8.1)

$$Tr(F_H^2)\wedge\omega^{m-2} \le -\frac{(2\pi)^2}{r}\left[(\Lambda\ Tr\ F_H)^2 - r\ Tr\ (\Lambda F_H)^2\right]\wedge\omega^m + \frac{1}{r}(Tr\ F_H)^2\wedge\omega^{m-2}$$

on M. Since f is nonpositive, it follows that

$$-f\ Tr(F_H^2)\wedge\omega^{m-2} \ge -f\left\{\frac{(2\pi)^2}{r}\left[(\Lambda\ Tr\ F_H)^2 - r\ Tr\ (\Lambda F_H)^2\right]\wedge\omega^m + \frac{1}{r}(Tr\ F_H)^2\wedge\omega^{m-2}\right\}.$$

Since the uniform norm of ΛF is assumed to be bounded by a constant, it follows that

$$-\int_M f\ Tr(F_H^2)\wedge\omega^{m-2}$$

is bounded from below by a constant. So we have $\frac{2\pi m}{\mu(m-1)}\mathcal{M}_{M'} \le \mathcal{M}_M$ + constant under the assumption that $\det H = \det H_0$ and the uniform norm of ΛF is bounded.

(5.4) For the proof of the existence of Hermitian-Einstein metrics for stable bundles by induction on the dimension of the base manifold, the Donaldson functional will be used to control the uniform norm of the Hermitian metric which is the solution of the heat equation. So we would like to know how to estimate a Hermitian metric in terms of the Donaldson functional when there is a Hermitian-Einstein metric.

Let H_0 be a Hermitian-Einstein metric of E. Any Hermitian metric is of the form $e^S H_0$ for some section S of End(E) over M which is Hermitian with respect to H_0, because any positive definite Hermitian matrix is the exponential of a Hermitian matrix. We would like to prove the following inequality

(5.4.1) $$\|\log \operatorname{Tr} e^S\|^2_{L^1(M)} \leq \text{Constant}\cdot\left[\mathcal{M}(H_0, e^S H_0) + \mathcal{M}(H_0, e^S H_0)^2\right].$$

To prove this inequality we join H_0 to $e^S H_0$ by the straight line-segment $e^{tS} H_0$ $(0 \leq t \leq 1)$. Let $h = H\,H_0^{-1} = e^{tS}$. The endomorphism h of E is Hermitian with respect to both H and H_0, *i.e.* $\langle hs, t\rangle_H = \langle s, ht\rangle_H$ and $\langle hs, t\rangle_{H_0} = \langle s, ht\rangle_{H_0}$. In terms of indices this means that both $h_\alpha^{\ \gamma}(H_0)_{\gamma\bar\beta}$ and $h_\alpha^{\ \gamma} H_{\gamma\bar\beta}$ are Hermitian matrices. We use the overhead dot to denote differentiation with respect to t. Since $\dot h h^{-1} = S$, it follows that $\frac{d\mathcal{M}}{dt} = \int_M \operatorname{Tr}((\Lambda F - \lambda I)\, S)\omega^m$ and

$$\frac{d^2\mathcal{M}}{dt^2} = \int_M \operatorname{Tr}(\Lambda(\frac{\partial F}{\partial t})\, S)\omega^m = \int_M i\ \operatorname{Tr}((\frac{\partial F}{\partial t})\, S)\omega^{m-1}.$$

Since H_0 is Hermitian-Einstein, we have $\Lambda F = \lambda I$ at $t = 0$ and $\frac{d\mathcal{M}}{dt} = 0$ at $t = 0$. From $\frac{\partial F}{\partial t} = \bar\partial(\partial_H(\dot h h^{-1})) = \bar\partial(\partial_H S)$ it follows that

$$\frac{d^2\mathcal{M}}{dt^2} = \int_M \sqrt{-1}\ \mathrm{Tr}((\bar\partial(\partial_H S))\ S)\omega^{m-1}$$
$$= \int_M \sqrt{-1}\ (\bar\partial(\partial_H S_\alpha{}^\beta))\ S_\beta{}^\alpha\ \omega^{m-1}$$
$$= \int_M \sqrt{-1}\ (\partial_H S_\alpha{}^\beta)\wedge\bar\partial S_\beta{}^\alpha\ \omega^{m-1}.$$

Now

$$\partial_H S_\alpha{}^\beta = \partial_H(H^{\beta\bar\sigma} H_{\tau\bar\sigma} S_\alpha{}^\tau) = H^{\beta\bar\sigma}\ \partial_H(H_{\tau\bar\sigma} S_\alpha{}^\tau)$$
$$= H^{\beta\bar\sigma}\ \overline{\bar\partial(\overline{H_{\tau\bar\sigma} S_\alpha{}^\tau})} = H^{\beta\bar\sigma}\ \overline{\bar\partial(H_{\tau\bar\alpha} S_\sigma{}^\tau)}$$

(because $H_{\tau\bar\sigma} S_\alpha{}^\tau$ is a Hermitian matrixs)

$$= H^{\beta\bar\sigma}\ H_{\tau\bar\alpha}\ \overline{\bar\partial S_\sigma{}^\tau}.$$

Hence

$$\frac{d^2\mathcal{M}}{dt^2} = \int_M \sqrt{-1}\ H^{\beta\bar\sigma}\ H_{\tau\bar\alpha}\ \overline{\bar\partial S_\sigma{}^\tau}\wedge\bar\partial\ S_\beta{}^\alpha\ \omega^{m-1} = m!\int_M \langle\bar\partial\ S,\bar\partial\ S\rangle_H.$$

We are going to integrate this twice with respect to t. We can do this by integrating the integrand $\langle\bar\partial\ S,\bar\partial\ S\rangle_H$ twice with respect to t at any given point P of M. At the point P we can choose a local trivialization of E so that at the ponit P the matrix H_0 is the identity matrix and S is diagonal with eigenvalues $\lambda_1,\cdots,\lambda_r$. Then

$$\langle\bar\partial\ S,\bar\partial\ S\rangle_H = \Sigma_{\alpha,\beta=1}^r\ |\bar\partial\ S_\alpha{}^\beta|\ \exp((\lambda_\beta - \lambda_\alpha)t).$$

Since $\mathcal{M}$ clearly vanishes at $t = 0$ by definition, it follows from $\frac{d\mathcal{M}}{dt} = 0$ and twice integrating $\frac{d^2\mathcal{M}}{dt^2}$ with respect to t that at $t = 1$ we have

$$\mathcal{M}(H_0, e^S H_0) = m! \int_M \Sigma^r_{\alpha,\beta=1} \, |\bar\partial \, S_\alpha{}^\beta|^2 \, \frac{e^{(\lambda_\beta - \lambda_\alpha)} - (\lambda_\beta - \lambda_\alpha) - 1}{(\lambda_\beta - \lambda_\alpha)^2},$$

where at every point P of M the integrand is calculated with respect to the trivialization of E so that H_0 is the identity matrix and S is diagonal with eigenvalues $\lambda_1, \cdots, \lambda_r$. We now use the inequality $e^x - x - 1 \geq \frac{x^2}{2}(1+x^2)^{-1/2}$ for all $x \in \mathbb{R}$. This simple inequality will be verified later. From

$$\frac{e^{(\lambda_\beta - \lambda_\alpha)} - (\lambda_\beta - \lambda_\alpha) - 1}{(\lambda_\beta - \lambda_\alpha)^2} \geq \frac{1}{2}\left[1 + (\lambda_\beta - \lambda_\alpha)^2\right]^{-1/2} \geq \frac{1}{2}\left[1 + 2\lambda_\beta^2 + 2\lambda_\alpha^2\right]^{-1/2}$$

it follows that

$$\Sigma^r_{\alpha,\beta=1} \, |\bar\partial \, S_\alpha{}^\beta| = \Sigma^r_{\alpha,\beta=1} \, |\bar\partial \, S_\alpha{}^\beta| \, 2^{-1/2}\left[1 + 2\lambda_\beta^2 + 2\lambda_\alpha^2\right]^{-1/4} \sqrt{2}\left[1 + 2\lambda_\beta^2 + 2\lambda_\alpha^2\right]^{1/4}$$

$$\leq \Sigma^r_{\alpha,\beta=1} \, |\bar\partial \, S_\alpha{}^\beta| \left[\frac{e^{(\lambda_\beta - \lambda_\alpha)} - (\lambda_\beta - \lambda_\alpha) - 1}{(\lambda_\beta - \lambda_\alpha)^2}\right]^{1/2} \sqrt{2}\left[1 + 2\lambda_\beta^2 + 2\lambda_\alpha^2\right]^{1/2}$$

$$\leq \left[\Sigma^r_{\alpha,\beta=1} \, |\bar\partial \, S_\alpha{}^\beta|^2 \, \frac{e^{(\lambda_\beta - \lambda_\alpha)} - (\lambda_\beta - \lambda_\alpha) - 1}{(\lambda_\beta - \lambda_\alpha)^2}\right]^{1/2} \left[\Sigma^r_{\alpha,\beta=1} \, 2(1 + 2\lambda_\beta^2 + 2\lambda_\alpha^2)\right]^{1/2}$$

$$= \left[\Sigma^r_{\alpha,\beta=1} \, |\bar\partial \, S_\alpha{}^\beta|^2 \, \frac{e^{(\lambda_\beta - \lambda_\alpha)} - (\lambda_\beta - \lambda_\alpha) - 1}{(\lambda_\beta - \lambda_\alpha)^2}\right]^{1/2} \left[2r^2 + 4r\Sigma^r_{\alpha,\beta=1} \, |S_\alpha{}^\beta|^2\right]^{1/2}$$

$$\leq \left[\Sigma^r_{\alpha,\beta=1} \, |\bar\partial \, S_\alpha{}^\beta|^2 \, \frac{e^{(\lambda_\beta - \lambda_\alpha)} - (\lambda_\beta - \lambda_\alpha) - 1}{(\lambda_\beta - \lambda_\alpha)^2}\right]^{1/2} \left[\sqrt{2}r + 2\sqrt{r} \, \Sigma^r_{\alpha,\beta=1} \, |S_\alpha{}^\beta|\right].$$

Thus

$$\left(\int_M |\bar\partial S|_{H_0}\right)^2 \leq c \, \mathcal{M}(H_0, H_0 e^S) \left[1 + \int_M |S|_{H_0}\right].$$

We claim that there exists a constant C such that $\|S\|_{L^1(M,H_0)} \leq C\,\|\bar\partial S\|_{L^1(M,H_0)}$ for any endomorphism S of E which is pointwise trace-free and which is Hermitian with respect to H_0. Suppose the contrary. Then we can find a sequence S_ν of such endomorphisms of E such that $\|S_\nu\|_{L^1(M,H_0)} = 1$ and $\|\bar\partial S_\nu\|_{L^1(M,H_0)} \to 0$ as $\nu \to \infty$. Since the Hermitian property of S_ν implies that $|\bar\partial S_\nu|^2_{H_0} = |\partial_{H_0} S_\nu|^2_{H_0}$, by Rellich's lemma there exists a subsequence of S_ν which converges to some S_∞ strongly in $L^1(M,H_0)$. So $\bar\partial S_\infty$ is zero in the sense of distributions and S_∞ is a holomorphic endomorphism of E. Since E is stable, every holomorphic endomorphism of E is a homothety and the torsion-free property of S_∞ implies that $S_\infty = 0$, contradicting the fact that $\|S_\infty\|_{L^1(M,H_0)}$ being the limit of $\|S_\nu\|_{L^1(M,H_0)}$ is 1. We now have

$$\|S\|^2_{L^1(M,H_0)} \leq C^2 c\, \mathcal{M}(H_0, e^S H_0) \left[1 + \|S\|_{L^1(M,H_0)}\right]$$

and

$$\|S\|^2_{L^1(M,H_0)} \leq \text{Constant}\cdot \left[\mathcal{M}(H_0, e^S H_0) + \mathcal{M}(H_0, e^S H_0)^2\right].$$

We have the inequality $|\log \operatorname{Tr} e^S| \leq \log r + |S|_{H_0}$, because from $|S|_{H_0} = \Sigma_{\nu=1}^r |\lambda_\nu|$ with $\lambda_1 \geq \cdots \geq \lambda_r$ and $\operatorname{Tr} e^S = \log\left[\Sigma_{\nu=1}^r e^{\lambda_\nu}\right]$ we have

$$-|S|_{H_0} \leq \lambda_r \leq \log \operatorname{Tr} e^S \leq \log(r\, e^{\lambda_1}) = \log r + \lambda_1 \leq \log r + |S|_{H_0}.$$

Hence

$$\| \log \text{ Tr } e^S \|^2_{L^1(M)} \leq \text{Constant} \cdot \left[\mathcal{M}(H_0, e^S H_0) + \mathcal{M}(H_0, e^S H_0)^2 \right].$$

We would like to remark that if there is a smooth family of M over an open neighborhood U of O in $\mathbb{R}^\ell$ and a smooth family of holomorphic stable vector bundles E over each member M of the family. If each M in the family carries a Hermitian–Einstein metric H_0 for the bundle E over it and H_0 is a smooth function of the variable of U, then after we shrink U we can assume that the constant C' can be chosen to be independent of the variable of U. The reason is that the only step that may give us trouble is the proof of $\|S\|_{L^1(M,H_0)} \leq C \|\bar{\partial}S\|_{L^1(M,H_0)}$ by the use of Rellich's lemma. When we have a parameter in U, we can still repeat the same argument with the S_υ for different υ possibly defined over a manifold M with different parameters.

We now verify the inequality $e^{-x} + x - 1 \geq \frac{x^2}{2}(1+x^2)^{-1/2}$. When $x \geq 0$ we clearly we have $e^x \geq 1 + x + \frac{x^2}{2}$. Replace x by $-x$. The inequality becomes $e^{-x} + x - 1 \geq \frac{x^2}{2}(1+x^2)^{-1/2}$. Now

$$e^{-x} + x - 1 \geq \frac{x^2}{2} - \frac{x^3}{6} + \frac{x^4}{24} - \frac{x^5}{120} = \frac{x^2}{2}\left[\left(1 - \frac{x}{3}\right) + \frac{x^2}{12}\left(1 - \frac{x}{5}\right) \right].$$

Since $\left(1 - \frac{x}{3}\right) + \frac{x^2}{12}\left(1 - \frac{x}{5}\right) \geq (1+x^2)^{-1/2}$ for $0 \leq x \leq 1$ as one can verify by observing that the left-hand side is decreasing and the right-hand side is increasing and the inequality holds at $x = 1$. The derivative of $e^{-x} + x - 1$ is $-e^{-x} + 1$ which is increasing. So the derivative of $e^{-x} + x - 1$ is at least $1 - \frac{1}{e}$ for $x \geq 1$. On the other hand the first derivative of $\frac{x^2}{2}(1+x^2)^{-1/2}$ is $\frac{x}{2}(x^2+2)(1+x^2)^{-3/2}$ and the first derivative of $\frac{x}{2}(x^2+2)(1+x^2)^{-3/2}$ is $\frac{1}{2}(2-x^2)(1+x^2)^{-5/2}$. So the maximum of

$\frac{x}{2}(x^2+2)(1+x^2)^{-3/2}$ for $x \geq 1$ is achieved at $x = \sqrt{2}$ and it is equal to $\left(\frac{2}{3}\right)^{3/2}$ which is less than $1 - \frac{1}{e}$. Thus $e^{-x} + x - 1 - \frac{x^2}{2}(1+x^2)^{-1/2}$ has positive derivative for $x \geq 1$ and is increasing for $x \geq 1$. Since $e^{-x} + x - 1 - \frac{x^2}{2}(1+x^2)^{-1/2}$ is positive at $x = 1$, it follows that it is positive for all $x \geq 1$. This concludes the proof of the inequality $e^{-x} + x - 1 \geq \frac{x^2}{2}(1+x^2)^{-1/2}$.

Now let $H = e^S H_0$ and $h = H H_0^{-1} = e^S$. We have by (1.9.2)

$$\Delta \log \operatorname{Tr} h \geq -(|\Lambda F_{H_0}| + |\Lambda F_H|).$$

Let $G(P,Q)$ be the Green's function of M for the Laplacian Δ. Then for any smooth function $\varphi(P)$ on M we have

$$\varphi(P) = \frac{1}{\operatorname{Vol} M}\int_{Q\in M} \varphi(Q)dQ - \int_{Q\in M} G(P,Q)(\Delta\varphi)(Q)dQ.$$

Let K be a positive number so that $-K$ is a lower bound for $G(P,Q)$. Then

$$\begin{aligned}\varphi(P) &= \frac{1}{\operatorname{Vol} M}\int_{Q\in M} \varphi(Q)dQ - \int_{Q\in M} (G(P,Q) + K)(\Delta\varphi)(Q)dQ \\ &\leq \frac{1}{\operatorname{Vol} M}\int_{Q\in M} \varphi(Q)dQ + \sup_M(-\Delta\varphi)\int_{Q\in M} (G(P,Q) + K)dQ \\ &= \frac{1}{\operatorname{Vol} M}\int_{Q\in M} \varphi(Q)dQ + K(\operatorname{Vol} M)\sup_M(-\Delta\varphi).\end{aligned}$$

We apply this to $\varphi = \log \operatorname{Tr} h$ and use (5.4.1) to get

$$\begin{aligned}(5.4.2)\quad \sup_M \log \operatorname{Tr} h &\leq \frac{1}{\operatorname{Vol} M}\|\log \operatorname{Tr} h\|_{L^1(M)} + K(\operatorname{Vol} M)\sup_M(-\Delta \log \operatorname{Tr} h) \\ &\leq \text{Constant}\cdot\left[\mathcal{M}(H_0,H)^{1/2} + \mathcal{M}(H_0,H) + \sup_M(|\Lambda F_{H_0}| + |\Lambda F_H|)\right].\end{aligned}$$

Again if there is a smooth family of M over an open neighborhood U of 0 in $\mathbb{R}^\ell$ and a smooth family of holomorphic stable vector bundles E over each member M of the family so that each M in the family carries a Hermitian-Einstein metric H_0 for the bundle E over it and H_0 is a smooth function of the variable of U, then after we shrink U we can assume that the constant in the inequality (5.4.2) can be chosen to be independent of the variable of U.

§6 *Convergence of Solution at Infinite Time*

(6.1) Now we are ready to handle the convergence of the solution H of the heat equation when t goes to infinity. We want to show that the limit of H exists as $t \to \infty$ and the limit is Hermitian-Einstein. We use induction on the complex dimension m of the base manifold. We assume that when the complex dimension of the base manifold is $m - 1 \geq 1$, every stable holomorphic vector bundle admits a Hermitian-Einstein metric. Now we prove the existence of Hermitian-Einstein metrics for stable bundles over a base manifold of complex dimension m.

We take a *generic* projective algebraic hypersurface M' of M in the class $\mu\,\omega$ for μ sufficiently large so that $E|M'$ is a stable bundle over M'. A proof will be given in Appendix B of this Chapter for the statement that the restriction of a stable bundle to a generic hypersurface cut out by a sufficiently high power of an ample divisor is again stable. We can assume that we have a holomorphic family of such M' parametrized by the open disc D_r in $\mathbb{C}$ of positive radius r so that the union of all M' in the family contains a nonempty open subset Ω of M. By induction hypothesis (after replacing r by a smaller one if necessary) we have a Hermitian-Einstein metric H_0' for each member $E|M'$ of the family parametrized by D_r so that H_0' is a smooth function of the variable of D_r. For the Donaldson function $\mathcal{M}_{M'}$ of M' we have the inequality for the $\frac{2\pi m}{\mu(m-1)}\,\mathcal{M}_{M'}(H_0,H) \leq \mathcal{M}_M(H_0,H)$ + constant, because $\det H = \det H_0$ and ΛF is uniformly bounded. The constant in the inequality is independent of the parameter in D_r after replacing r by a smaller one if necessary. Since $\mathcal{M}_M(H_0,H)$ is a nonincreasing function of t, it follows that $\mathcal{M}_{M'}(H_0,H)$ is bounded from above by a constant independent of t and independent of the parameter in D_r. Since $\mathcal{M}_{M'}(H_0',H) \leq \mathcal{M}_{M'}(H_0',H_0) + \mathcal{M}_{M'}(H_0,H)$, it follows that $\mathcal{M}_{M'}(H_0',H)$ is also bounded from above by a constant independent of t and independent of the parameter in D_r.

From the upper bound of $\mathcal{M}_{M'}(H_0',H)$ by (5.4.1) we know that $\|\log \operatorname{Tr} H(H_0')^{-1}\|_{L^1(M')}$ is bounded by a constant independent of the variable of D_r (after replacing r by a smaller one if necessary). It follows that $\|\log \operatorname{Tr} H(H_0)^{-1}\|_{L^1(M')}$ also is bounded by a constant independent of the variable of D_r. Hence $\|\log \operatorname{Tr} H(H_0)^{-1}\|_{L^1(\Omega)}$ is bounded by a constant independent of t (with some shrinking of Ω if necessary to avoid the singular points of the family M'). We have by (1.9.2)

$$\Delta \log \operatorname{Tr} h \geq -(|\Lambda F_{H_0}| + |\Lambda F_H|).$$

So $\Delta \log \operatorname{Tr} h$ is bounded from below uniformly in t. Fix a point P_0 in Ω and let $B_\delta = B_\delta(P_0)$ be a small closed geodesic ball of positive radius δ centered at P_0. Let $G_\delta(P,Q)$ $(P \in M - B_\delta,\ Q \in \partial B_\delta)$ be the Green's function for $M - B_\delta$ so that $G_\delta(P,Q) = 0$ for $Q \in \partial B_\delta$ and $\Delta_Q G_\delta(P,Q)$ is the delta function at P. Let $\Gamma_\delta(P,Q)$ be the boundary normal derivative of $G_\delta(P,Q)$ with respect to Q in the outward normal direction of $M - B_\delta$. Then we have the Green's formula

$$\varphi(P) = \int_{\partial B_\delta} \Gamma_\delta(P,Q)\varphi(Q)dQ + \int_{M-B_\delta} G_\delta(P,Q)\Delta\varphi(Q)dQ$$

for any function φ smooth on the topological closure of $M - B_\delta$. Suppose $B_{2\eta}(P_0)$ is contained in Ω. Apply the above Green's formula to $\log \operatorname{Tr} h$

and average over δ from $\frac{\eta}{2}$ to 2η. Since $G_\delta(P,Q)$ is nonpositive, from the bound of $\|\log \operatorname{Tr} h\|_{L^1(\Omega)}$ and the lower bound of $\Lambda \log \operatorname{Tr} h$ we obtain an upper bound of $\log \operatorname{Tr} h$ on $M - B_{2\eta}(P_0)$ independent of t. By applying the above argument to another point P_0' of Ω instead of P_0 so that $B_{2\eta}(P_0)$ and $B_{2\eta}(P_0')$ are disjoint subsets of Ω, we obtain an upper bound of $\log \operatorname{Tr} h$ on M independent of t. Since the determinant of h is 1, it follows that each eigenvalue of h is bounded from below on M by a positive number independent of t.

We would like to show that the L_1^2 norm of h is bounded by a positive number independent of t. Choose a finite number of smooth Hermitian metrics $H^{(\nu)}$, $1 \le \nu \le k$, along the fibers of E so that for any smooth Hermitian metric K along the fibers of E, with respect to a local trivialization of E, the entries of the Hermitian matrix representing K are linear functions of $\operatorname{Tr}(K(H^{(\nu)})^{-1})$, $1 \le \nu \le k$, whose coefficients are smooth functions depending only on $H^{(\nu)}$, $1 \le \nu \le k$. So the L_1^2 norm of H_t is bounded independent of t if the L_1^2 norm of $\operatorname{Tr}(H_t(H^{(\nu)})^{-1})$ is bounded independent of t for $1 \le \nu \le k$. We have by (1.9.3)

$$\Lambda \log \operatorname{Tr}\,(H_t(H^{(\nu)})^{-1}) \ge -(|\Lambda F_{H^{(\nu)}}| + |\Lambda F_{H_t}|)$$

from which we get

$$\Lambda \operatorname{Tr}\,(H_t(H^{(\nu)})^{-1}) \ge -(|\Lambda F_{H^{(\nu)}}| + |\Lambda F_{H_t}|)\operatorname{Tr}\,(H_t(H^{(\nu)})^{-1}).$$

Multiplying both sides by $\operatorname{Tr}\,(H_t(H^{(\nu)})^{-1})$ and integrating by parts we get

$$\int_M \left|\nabla \operatorname{Tr}\,(H_t(H^{(\nu)})^{-1})\right|^2 \le \int_M (|\Lambda F_{H^{(\nu)}}| + |\Lambda F_{H_t}|)\left|\operatorname{Tr}\,(H_t(H^{(\nu)})^{-1}).\right.$$

It follows from the supremum bounds of h and ΛF_{H_t} which are independent of t that the L_1^2 norm $\mathrm{Tr}\,(H_t(H^{(\nu)})^{-1})$ is independent of t. So the L_1^2 norm of H_t is bounded independent of t and the L_1^2 norm of h is bounded independent of t.

By the argument for the case $T = \infty$ in the last part of (3.5) we conclude from the uniform bound of ΛF_{H_t} for all t that for any finite p we have an L_2^p bound and a C^1 bound of H_t independent of t.

(6.2) We now verify that $\mathcal{M}_M(H_0,H_t)$ is bounded uniformly in t. To calculate $\mathcal{M}_M(H_0,H_t)$ we join H_0 to H_t by a path $K_{s,t} = (1-s)H_0 + s\,H_t$ for $0 \leq s \leq 1$. The curvature tensor $F_{K_{s,t}} = \bar\partial((\partial\, K_{s,t})K_{s,t}^{-1})$ of $K_{s,t}$ is bounded in L^p norm uniformly in t and s, because H_t^{-1} is bounded in C^0 and H_t is bounded in L_2^p norm uniformly in t. The secondary characteristic class $R_1(H_0,H_t) = \log\det(H_tH_0^{-1})$ is bounded in C^0 norm uniformly in t and the secondary characteristic class $R_2(H_0,H_t) = \sqrt{-1}\int_{s=0}^{1}\mathrm{Tr}(F_{K_{s,t}}\dot H_tK_{s,t}^{-1})ds$ is bounded in L^p norm uniformly in t for all p. Hence $\mathcal{M}(H_0,H_t) = \int_M m\,R_2(H_0,H_t)\Lambda\omega^{m-1} - \lambda R_1(H_0,H_t)\omega^m$ is bounded uniformly in t.

For our flow $\dot hh^{-1} = -(\Lambda F - \lambda I)$ we have

$$\frac{d\mathcal{M}}{dt} = -\int_M \mathrm{Tr}((\Lambda F - \lambda I)\wedge(\Lambda F - \lambda I))\omega^m$$
$$= -\int_M |\Lambda F - \lambda I|^2 \leq 0.$$

Also along our flow $\dot hh^{-1} = -(\Lambda F - \lambda I)$ we have

$$\frac{d^2\mathcal{M}}{dt^2} = -\int_M \frac{\partial}{\partial t}|\Lambda F - \lambda I|^2$$
$$= -2\,\mathrm{Re}\int_M \langle \Lambda\dot{F}, \Lambda F - \lambda I\rangle$$
$$= -2\,\mathrm{Re}\int_M \langle \Delta\Lambda F, \Lambda F - \lambda I\rangle$$
$$= 2\int_M \|\nabla\Lambda F\|^2 \geq 0.$$

Thus $\mathcal{M}(H_0, H_t)$ is a nonincreasing convex function of t.

(6.3) We make the following simple observation in calculus. If a nonincreasing convex function $f(t)$ of t is bounded from below as $t \to \infty$, its derivative $f'(t)$ must go to 0 as $t \to \infty$. The reason is as follows. Let b be its lower bound. For any fixed t_0. By the mean value theorem, for any $t > t_0$ the difference quotient $\dfrac{f(t) - f(t_0)}{t - t_0}$ is equal to some $f'(t_1)$ with $t \leq t_1 \leq t$. By the convexity of f we know that f' is increasing and $f'(t) \geq f'(t_1)$. Since f is nonincreasing, it follows that $f'(t)$ is nonpositive. Hence

$$0 \geq f'(t) \geq \frac{f(t) - f(t_0)}{t - t_0} \geq \frac{b - f(t_0)}{t - t_0} \to 0$$

as $t \to \infty$.

Now we apply this simple calculus observation to the function $f(t) = \mathcal{M}(H_0, H_t)$. We conclude that the limit of $\dfrac{d\mathcal{M}}{dt} = -\int_M |\Lambda F_{H_t} - \lambda I|^2$ is zero as $t \to \infty$. This means that $\Lambda F_{H_t} - \lambda I$ converges strongly in L^2 to zero as $t \to \infty$.

(6.4) Since we have an L_2^p bound and a C^1 bound of H_t (and H_t^{-1}) independent of t. We can select a subsequence t_i going to infinity such that H_{t_i} approaches a Hermitian metric H_∞ of E in C^0 norm. By replacing t_i by a subsequence we can assume that H_{t_i} approaches a weak limit in L^p_2. This weak limit must be equal to H_∞. Hence H_∞ is in L^p_2. By using Rellich's lemma and replacing t_i by a subsequence we can assume also that H_{t_i} approaches H_∞ in L^p_1 for all p. Hence $\Lambda F_{H_{t_i}} - \lambda I$ approaches $\Lambda F_{H_\infty} - \lambda I$ weakly in L^p for all p as $i \to \infty$. Since $\Lambda F_{H_t} - \lambda I$ approaches zero strongly in L^2 as $t \to \infty$, it follows that $\Lambda F_{H_\infty} = \lambda I$, from which we conclude by standard elliptic estimate arguments that H_∞ is smooth and is Hermitian-Einstein. More precisely $\Lambda F_{H_\infty} = \lambda I$ means $\Delta H_\infty = \Lambda(\partial H_\infty)H_\infty(\bar{\partial} H_\infty) + \lambda H_\infty$ (when with respect to a holomorphic local trivialization of E the metric H_∞ is regarded as a Hermitian matrix). From the L^p_k bound of H_∞ for all p we obtain the L^p_{k-1} bound of $\Lambda(\partial H_\infty)H_\infty(\bar{\partial} H_\infty) + \lambda H_\infty$ for all p. Using the ellipticity of Δ, we get the L^p_{k+1} bound of H_∞. So by induction on k we get the smoothness of H_∞. Note that to argue rigorously one should take a partition of unity $\{\rho_i\}$ of M subordinate to an open cover of M so that E is holomorphically trivial over each member of the open cover and apply the above argument to the equation for $\Delta(\rho_i H_\infty)$ instead of to the equation for ΔH_∞. This concludes the proof of the existence of Hermitian-Einstein metrics for stable bundles by induction on the dimension of the base manifold when one assumes as proved the case of the base manifold being a complex curve. The case of the base manifold being a complex curve was first proved by Narasimhan and Seshadri [N-S] with an alternative proof given later by Donaldson [D1]. We will give a proof of the case of the base manifold being a complex curve in Appendix A of this Chapter.

(6.5) *Remark.* If one does not use the argument of the last part of (3.5), one can still select in the following way a subsequence t_ν of t

approaching infinity so that H_{t_ν} converges weakly in L_2^2 and strongly in L_1^2 to some H_∞ so that $\Lambda F_{H_\infty} = \lambda I$.

From the L_1^2 bound of H_t and the L^2 bound of F_{H_t} and the equation $F_{H_t} = \bar{\partial}((\partial H_t)H_t^{-1})$ we conclude that the L_2^2 norm of H_t is independent of t. The L_2^2 norm of H_t^{-1} is also bounded independent of t, because each eigenvalue of h is bounded from below on M by a positive number independent of t. We can select a subsequence t_ν of t approaching infinity so that H_{t_ν} converges weakly in L_2^2 and strongly in L_1^2 to some H_∞ and $H_{t_\nu}^{-1}$ converges strongly in L^2 to some K_∞ and $(\partial H_{t_\nu})H_{t_\nu}^{-1}$ converges strongly in L^2 to some R_∞ as $\nu \to \infty$. Since $I = H_{t_\nu}H_{t_\nu}^{-1}$ converges strongly in L^1 to $H_\infty K_\infty$, it follows that $K_\infty = H_\infty^{-1}$ almost everywhere and H_∞^{-1} exists almost everywhere. Since $\partial H_{t_\nu} = \left[(\partial H_{t_\nu})H_{t_\nu}^{-1}\right]H_{t_\nu}$ converges strongly in L^1 to $R_\infty H_\infty$ and ∂H_∞ at the same time, we have $R_\infty = (\partial H_\infty)H_\infty^{-1}$ almost everywhere. So we conclude that $(\partial H_{t_\nu})H_{t_\nu}^{-1}$ converges strongly in L^2 to $(\partial H_\infty)H_\infty^{-1}$. Thus $(\partial H_{t_\nu})H_{t_\nu}^{-1}(\bar{\partial}H_{t_\nu})$ converges strongly in L^1 to $(\partial H_\infty)H_\infty^{-1}(\bar{\partial}H_\infty)$. Finally $F_{H_{t_\nu}} = -\bar{\partial}\partial H_{t_\nu} + (\partial H_{t_\nu})H_{t_\nu}^{-1}(\bar{\partial}H_{t_\nu})$ converges weakly in L^1 to $F_{H_\infty} = -\bar{\partial}\partial H_\infty + (\partial H_\infty)H_\infty^{-1}(\bar{\partial}H_\infty)$ as $\nu \to \infty$ and $\Lambda F_{H_{t_\nu}}$ converges weakly in L^1 to ΛF_{H_∞} as $\nu \to \infty$. We know from (6.3) that $\Lambda F_{H_t} - \lambda I$ converges strongly in L^2 to zero as $t \to \infty$. Hence $\Lambda F_{H_\infty} - \lambda I = 0$.

However I would like to point out that when one has only the L_2^2 bound and the uniform bound of H_∞ and the uniform positive lower bound of each of its eigenvalues, it is by no means clear that one can use standard elliptic estimate arguments to conclude from $\Lambda F_{H_\infty} = \lambda I_\infty$ that H_∞ is smooth. The difficulty is the term $\Lambda(\partial H_\infty)H_\infty(\bar\partial H_\infty)$ in the form $\Delta H_\infty = \Lambda(\partial H_\infty)H_\infty(\bar\partial H_\infty) + \lambda H_\infty$ of the equation $\Lambda F_{H_\infty} = \lambda I_\infty$.

APPENDIX A. Hermitian-Einstein Metrics for Stable Bundles Over Curves

(A.1) *The Choice of a Good Gauge.* To prove the existence of Hermitian-Einstein metrics for stable bundles over curves, we use the method of minimizing the global norm of the L^2 of the curvature tensor. The main problem is to prove that the minimum is achieved by some metric which will be the sought-after Hermitian-Einstein metric. There are two ways of looking at the limit of a minimizing sequence. We will use the second way. The first way is to fix a local basis of E and see whether the metric $H(t)$ when expressed in terms of the local basis converges to a *nonsingular* metric. The second way is to choose for every t a *suitable* local basis of E which is *unitary* with respect to the metric $H(t)$ and express the complex metric connection of $H(t)$ as a matrix-valued 1-form and see whether this matrix-valued 1-form converges. We call a local unitary basis of E a *gauge* for the connection. The problem of choosing suitable local unitary bases is the problem of choosing a good gauge. This was first done by Uhlenbeck [U] and the method is to use the implicit function theorem. (The L^p estimate for the best p was pointed out by Taubes [U, p.31]. Actually we will need only a very weak result from the result of choosing a good gauge.) When we have the convergence of the matrix-valued 1-form, we can define a complex structure on E by characterizing local holomorphic sections of the new complex structure as those sections whose (0,1)-directive with respect to the limit connection (given by the limit matrix-valued 1-form) is zero. The second way enables us to use the condition of stability. Stability will imply that the new complex structure of E must be biholomorphic to the original complex structure of E. The local basis of E with respect to which the matrix-valued 1-form defining the new complex structure is represented gives us a Hermitian metric. This metric will be shown to be the Hermitian-Einstein metric we need.

(A.2) The problem of choosing a good gauge is a local problem. We look at the case of a trival bundle over a complex Euclidean domain. Suppose we have a connection for a trivial holomorphic Hermitian vector bundle over a domain U in $\mathbb{C}^m$ with curvature F. On U we have a Kähler metric $g_{\alpha\bar{\beta}}$. We assume that we have a bound on ΛF which is the contraction of F by $g_{\alpha\bar{\beta}}$.

By a *good gauge* we mean a smooth basis of the vector bundle so that (i) the connection A with respect to the smooth basis satisfies $d^*A = 0$ and (ii) A vanishes at the normal direction at every point of the boundary. Here d^* means the *formal* adjoint of the operator d. Before we use the implicit function theorem to show the existence of such a good gauge, we first explain why we want such a gauge. The reason for the choice of such a gauge is that with respect to such a gauge we can get bounds on A from bounds on the curvature tensor F, because $dA - A\wedge A = F$ and $d^*A = 0$ together form an *elliptic system* and the boundary condition for that system is the vanishing of A at the boundary normals.

(A.3) To prove the existence of a good gauge, we look at the case of the open unit ball B in $\mathbb{C}^m$ and assume that we have a smooth connection A whose curvature F_A has L^m norm less than a fixed small positive number κ.

We would like to find a gauge s so that the equations $d^*A = 0$ and $A(\text{boundary normal}) = 0$ are satisfied. First we choose a gauge by taking a unitary frame and integrate along radial lines by parallel transport. This particular gauge is known as the *exponential gauge*. With respect to this exponential gauge the connection A becomes a matrix valued 1-form. We will use the notation to denote both the connection and associated matrix valued 1-form with respect to a gauge. Which one is meant will be clear from the context. We pull back A by the dilation map $x \to tx$ $(0 \leq t \leq 1)$ and thereby get a path of connections A_t joing A to the trivial connection.

The L^m norm on B of the curvature F_{A_t} of A_t is no more than the L^m norm of the curvature F_A of A on the ball of radius t. Moreover, A_t vanishes at the boundary normal of B.

We are going to use the method of continuity to choose a gauge given by the unitary frame s_t so that the new connections $\tilde{A}_t$ with respect to the gauge s_t satisfy the equations $d^*\tilde{A}_t = 0$ and $\tilde{A}_t(\text{boundary normal}) = 0$ are satisfied. Moreover, we require that the L_1^m norm of $\tilde{A}_t$ is less than $c\kappa$, where c is some univeral constant to be specified later. We denote also by s_t the unitary matrix expressing the gauge s in terms of the exponential gauge.

Closedness can be seen in the following way. Suppose we consider $t < t_*$. We have the equations

$$d^*\tilde{A}_t = 0, \quad d\tilde{A}_t - \tilde{A}_t \wedge \tilde{A}_t = s_t F s_t^{-1}, \quad \tilde{A}_t(\text{boundary normal}) = 0.$$

The key point is that the supremum of the L_1^m norm of the connection $\tilde{A}_t$ must be still *less than* $c\kappa$ instead of *equal* to it. Now $d^*\tilde{A}_t$ together with $d\tilde{A}_t$ forms an elliptic system and s_t is unitary. So

$$\|\tilde{A}_t\|_{L_1^m} \leq C(\|F\|_{L^m} + \|\tilde{A}_t \wedge \tilde{A}_t\|_{L^m}) \leq C\kappa + C'\|\tilde{A}_t\|^2_{L^{2m}},$$

but by Sobolev lemma in real dimension $2m$ one has $\|\tilde{A}_t\|_{L^{2m}} \leq C''\|\tilde{A}_t\|_{L_1^m}$. Here C and C' and other similar constants introduced later are universal constants. Hence

$$\|\tilde{A}_t\|_{L_1^m} \leq C\kappa + C'C''\|\tilde{A}_t\|^2_{L_1^m} \leq C\kappa + C'C''c^2\kappa^2$$

and when $C + C'C''c^2\kappa < c$ we have the conclusion we want. Thus it suffices to choose $c = 2C$ and $\kappa < \frac{1}{4CC'C''}$.

We have $ds_t s_t^{-1} + s_t A_t s_t^{-1} = \tilde{A}_t$. Thus we have $\|s_t\|_{L_2^m} \leq C^{\#}$, because $\|\tilde{A}_t\|_{L_1^m} \leq c\kappa$ and A_t is smooth in the space variable as well as the variable t. Hence for any small positive ϵ we can select a subsequence t_ν approaching t_* from the left so that s_{t_ν} converges in $L_{2-\epsilon}^m$ to some s_{t_*} which is in L_2^m. We now define $\tilde{A}_{t_*}$ by $ds_{t_*} s_{t_*}^{-1} + s_{t_*} A_t s_{t_*}^{-1}$ and we have closedness, because by Fatou's lemma clearly $\|\tilde{A}_{t_*}\|_{L_1^m}$ is dominated by the supremum of $\|\tilde{A}_t\|_{L_1^m}$ for $t < t_*$.

(A.4) Now we look at openness. Suppose we have a solution s_{t_*} for some $t_* < 1$. Let $\tilde{s}_t = s_{t_*}^{-1} s_t$ and $\hat{A}_t = ds_{t_*} s_{t_*}^{-1} + s_{t_*} A_t s_{t_*}^{-1}$. Consider the equation

$$d^*(d\tilde{s}_t \tilde{s}_t^{-1} + \tilde{s}_t \hat{A}_t \tilde{s}_t^{-1}) = 0$$

near $t = t_*$. At $t = t_*$ the solution is $\tilde{s} = I$. We take the linearized equation at $t = t_*$ and get

$$d^* d\dot{s} + \langle d\dot{s}, \hat{A}_t \rangle - \langle \hat{A}_t, d\dot{s} \rangle = -d^* \dot{A}.$$

The adjoint of the operator $\dot{s} \to d^* d\dot{s} + \langle d\dot{s}, \hat{A} \rangle - \langle \hat{A}, d\dot{s} \rangle$ is $\dot{s} \to d^* d\dot{s} + \langle d\dot{s}, \hat{A}_t \rangle - \langle \hat{A}, d\dot{s}_t \rangle$ which is injective, as one can see by using the Sobolev estimates

$$\|\dot{s}\|_{L_2^m} \le C\|\dot{s}\|_{L_1^m}\|\hat{A}_t\|_{L^{2m}}$$
$$\le C'\|\dot{s}\|_{L_2^m}\|\hat{A}_t\|_{L_1^m}$$

and by setting at the very beginning $C'c\kappa < 1$ (recall that $d\tilde{s}_t$ vanishes at the boundary normals). Hence we can get solutions $\tilde{s}_t$ for t sufficiently close to t_*. Let $s_t = s_{t_*}\tilde{s}_t$. Then we get solutions s_t for $d^*(ds_t s_t^{-1} + s_t A_t s_t^{-1}) = 0$.

We would like to remark that the above argument works even better when the $L^{m'}$ norm of F is bounded by a small universal constant depending on m' if m' is greater than m. In that case we get a good gauge with a bound on the $L_1^{m'}$ norm of A.

(A.5) Now we are ready to prove the existence of Hermitian-Einstein metrics for stable bundles over curves. Our compact complex manifold M is of complex dimension one. We use our old notation of denoting the stable holomorphic vector bundle of rank r by E. We consider only Hermitian metrics along the fibers of E with the property that the first Chern form of the complex metric connection is harmonic. Take a Hermitian metric H along the fibers of E and let $A = A_H$ be its complex metric connection and let $F = F_A = F_H$ be the curvature form of A. Before we take a sequence of A minimizing the global L^2 norm of F (or equivalently the global L^2 norm of the trace-free part of F), we first consider a sequence of complex Hermitian connections A_i of Hermitian metrics H_i of E with the global L^2 norm of F_{A_i} is bounded by some constant C independent of i.

The requirement for the existence of good gauges is that for some $p \ge 1$ the global L^p norm of the curvature tensor is smaller than some universal number depending on p. Since for a coordinate change $z = \epsilon\zeta$ we have $F_{\alpha\, 1\bar{1}}^{\ \beta}(z)dz\wedge d\bar{z} = \epsilon^2 F_{\alpha\, 1\bar{1}}^{\ \beta}(\epsilon\zeta)d\zeta\wedge d\bar{\zeta}$ it follows that the L^2 norm of F on

$\{|\zeta| < 1\}$ with respect to the coordinte ζ is equal to ϵ times the L^2 norm of F on $\{|z| < \epsilon\}$ with respect to the coordinate z. Hence we can cover M by a finite number of unit coordinate disks U_j so that the L^2 norm of F_{A_i} on each U_j with respect to the coordinate of U_j is less than the required universal constant. So we have a good gauge for $A_i^{(j)}$ over U_j with L_1^2 bound. In other words over each U_j we have a unitary frame field $e_{i\alpha}^{(j)}$, $1 \leq \alpha \leq r$, of E so that with respect to the frame field $e_{i\alpha}^{(j)}$, $1 \leq \alpha \leq r$, and the coordinate of U_j the L_1^2 norm on U_i of the $r \times r$ matrix valued 1-form $A_i^{(j)} = (A_i^{(j)}{}_\alpha^\beta)$ of the metric complex connection of the Hermitian metric H_i is uniformly bounded independent of i and j.

On $U_j \cap U_\ell$ let $g_i(j,\ell)$ with entries $g_i(j,\ell)_\alpha{}^\beta$ denote the unitary $r \times r$ matrix-valued function which relates the unitary frame fields $e_{i\alpha}^{(j)}$ and $e_{i\beta}^{(\ell)}$ on U_j and U_ℓ respectively, *i.e.* the equation $e_{i\alpha}^{(j)} = \Sigma_{\beta=1}^r g_i(j,\ell)_\alpha{}^\beta e_{i\beta}^{(\ell)}$ holds on $U_j \cap U_\ell$. It follows from the equation $dg_i(j,\ell) = A_i^{(j)} g_i(j,\ell) - g_i(j,\ell) A_i^{(\ell)}$ on $U_j \cap U_\ell$ and repeated applications of the Sobolev lemma that the global L_2^2 norm of $g_i(j,\ell)$ on $U_j \cap U_\ell$ is uniformly bounded in i.

By going to a subsequence if necessary, we can assume that $A_i^{(j)}$ converges weakly in L_1^2 and strongly in L^p for any finite p to some $A_\infty^{(j)}$. We can also assume that $g_i(j,\ell)$ converges weakly to some $g_\infty(j,\ell)$ in L_2^2 and the convergence is strong in L_1^p norm for any finite p. The transition functions $g_\infty(j,\ell)$ define a continuous vector bundle E_∞ by patching together trivial bundles over U_j. We denote the standard basis of the trivial bundle over U_j by $e_{\infty\alpha}^{(j)}$, $1 \leq \alpha \leq r$. The bundle E_∞ is topologically the same as E. Since each $g_\infty(j,\ell)$ is unitary at every point, we have a Hermitian metric along the fibers of E_∞. The connection $A_\infty^{(j)}$ for E_∞ is compatible with the Hermitian metric of E_∞. For convenience when we want to emphasize that the unitary frame fields $e_{i\alpha}^{(j)}$ and the matrix valued 1-form $A_i^{(j)}$ are being used, we denote E by E_i.

We would like to point out is that though the transition functions $g_i(j,\ell)$ approach $g_\infty(j,\ell)$, yet in general it is not true that in E the unitary frame fields $e_{i\alpha}^{(j)}$ approach some finite limit as $i \to \infty$. This difficulty is precisely the difficulty that the sequence of metrics H_i may blow up in some directions as $i \to \infty$. It makes no sense to say that the unitary frame fields $e_{i\alpha}^{(j)}$ of E_i approach the unitary frame fields $e_{\infty\alpha}^{(j)}$ of E_∞, unless we are doing the comparison in some fixed bundle $\tilde{E}$ using isomorphisms between E_i and $\tilde{E}$ and an isomorphism between E_∞ and $\tilde{E}$.

(A.6) Now we try to integrate the complex structure defined by the (0,1) component of the connection $A_\infty = (A_\infty^{(j)})$. One needs only do this locally. So we choose some U_j and express everything in terms of the unitary frame $e_{\infty\alpha}^{(j)}$. Sections of E now become r-vector valued functions and the connection is represented as an r×r matrix valued 1-form. We suppress now the superscript (j).

What we have to do is to locally solve the differential equation $\bar{\partial}_{A_\infty} f = 0$ for local r-vector-valued function f, where $\bar{\partial}_{A_\infty} f$ means $\bar{\partial} f + A_\infty^{(0,1)} f$. Here $A_\infty^{(0,1)}$ is an r×r matrix-valued (0,1)-form and $\bar{\partial}$ means the usual $\bar{\partial}$ operator of M applied to an r-tuple of functions. We have to get enough such solutions locally so that they can locally generate the vector bundle E at every point over $\mathbb{C}$. We use the standard technique of locally taking a section $\tilde{\varphi}$ of E whose $\bar{\partial}_{A_\infty}$ may not be zero. Then we form $\varphi = \bar{\partial}_{A_\infty} \tilde{\varphi}$ and try to solve $\bar{\partial}_{A_\infty} \psi = \varphi$ for ψ so that $\tilde{\varphi} - \psi$ would satisfy $\bar{\partial}_{A_\infty}(\tilde{\varphi} - \psi) = 0$. We would choose $\tilde{\varphi}$ with $\bar{\partial}_{A_\infty}\tilde{\varphi}$ small and conclude that we have a small solution ψ so that $\tilde{\varphi} - \psi$ differs not much from $\tilde{\varphi}$. When we start with a collection of $\tilde{\varphi}$ which generate E locally, our new collection $\tilde{\varphi} - \psi$ would also generate E locally. Now let us make this quantitatively precise.

Take a unit coordinate disk U of M with holomoprhic coordinate z. Let $\Psi(z) = A_\infty(\frac{\partial}{\partial \bar{z}})$. That is $A_\infty^{(0,1)} = \Psi(z)d\bar{z}$. When we do the coordinate change $z = \epsilon\zeta$, we have $A_\infty^{(0,1)} = \epsilon\Psi(\epsilon\zeta)d\bar{\zeta}$ and the L^p norm of $A_\infty^{(0,1)}$ on the disk $\{|\zeta| < 1\}$ with respect to the standard metric of $\{|\zeta| < 1\}$ is no more than $\epsilon^{\frac{p-2}{p}}$ times the L^p norm of $A_\infty^{(0,1)}$ on the disk $\{|z| < 1\}$ with respect to the standard metric of $\{|z| < 1\}$, because

$$\left[\iint_{|\zeta|<1} |\epsilon\Psi(\epsilon\zeta)|^p \sqrt{-1}\, d\zeta \wedge d\bar{\zeta}\right]^{1/p} = \epsilon^{\frac{p-2}{p}} \left[\iint_{|z|<\epsilon} |\Psi(z)|^p \sqrt{-1}\, dz \wedge d\bar{z}\right]^{1/p}.$$

So by replacing U by a smaller disk and then changing the coordinate to make the smaller disk again U we can assume that our local coodinate chart is the unit disc U in $\mathbb{C}$ and the L^p norm of $A_\infty^{(0,1)}$ on the unit disc U is les than a prescribed small positive number ϵ. In other words, the L^p norm of Ψ is less than ϵ. We rewrite the equation $\bar{\partial}_{A_\infty}\psi = \varphi$ as $\frac{\partial}{\partial \bar{z}}\psi(z) + \Psi(z)\psi(z) = \varphi(z)$.

Let φ be an r-tuple of functions. We would like to solve the equation $\frac{\partial}{\partial \bar{z}} f(z) + \Psi(z)f(z) = \varphi$ with estimates on f in terms of φ. First we solve $\frac{\partial}{\partial \bar{z}} f_0 = \varphi_0$. Let K be the L^p norm of f_0. We are going to solve inductively the equation $\bar{\partial} f_{\nu+1} = -\Psi f_\nu$ for $\nu \geq 0$ so that $\|f_\nu\|_{L^p} \leq C^\nu \epsilon^\nu K$, where C is a constant which we will later specify. To solve a differential equation $\frac{\partial}{\partial \bar{z}} u(z) = v(z)$, we consider the equation $\frac{\partial^2}{\partial z \partial \bar{z}} w(z) = v(z)$ and solve for $w(z)$ in terms of $w(z)$ by using the Newtonian potential and finally set $u(z) = \frac{\partial}{\partial z} w(z)$. We have the estimate $\|u\|_{L_1^q} \le C_q \|v\|_{L^q}$ where C_q is a universal constant (see *e.g.* [G-T, p. 230]). From the induction hypothesis we have $\|\Psi f_\upsilon\|_{L^{p/2}} \le \|\Psi\|_{L^p} \|f_\upsilon\|_{L^p} \le \epsilon C^\upsilon \epsilon^\upsilon K$. We solve $\frac{\partial}{\partial \bar{z}} f_{\upsilon+1} = - \Psi f_\upsilon$ for $f_{\upsilon+1}$ and get $\|f_{\upsilon+1}\|_{L_1^{p/2}} \le C_{p/2} \epsilon C^\upsilon \epsilon^\upsilon K$. By the Sobolev lemma we have $\|f_{\upsilon+1}\|_{L^p} \le C'_p \|f_{\upsilon+1}\|_{L_1^{p/2}} \le C'_p C_{p/2} \epsilon C^\upsilon \epsilon^\upsilon K$, where C'_p is a universal constant. The construction by induction is complete after we set $C = C'_p C_{p/2}$. We choose ϵ so that $C\epsilon < 1$. Let $f = \Sigma_{\upsilon=0}^\infty f_\upsilon$. Then $\frac{\partial}{\partial \bar{z}} f(z) + \Psi(z) f(z) = f$ from summing the equations $\frac{\partial}{\partial \bar{z}} f_{\upsilon+1} = - \Psi f_\upsilon$ for all $\upsilon \ge 0$ and using $\bar{\partial} f_0 = \varphi$. From $\|f_{\upsilon+1}\|_{L_1^{p/2}} \le C_{p/2} \epsilon C^\upsilon \epsilon^\upsilon K$ for $\upsilon \ge 0$ we have $\|\Sigma_{\upsilon=1}^\infty f_\upsilon\|_{L_1^{p/2}} \le \frac{C_{p/2} \epsilon K}{1 - C\epsilon}$. On the other hand we can solve $\frac{\partial}{\partial \bar{z}} f_0 = \varphi_0$ so that $\|f_0\|_{L_1^{p/2}} \le C_{p/2} \|\varphi\|_{L^{p/2}}$. Thus $\|f\|_{L_1^{p/2}} \le C_p^* \|\varphi\|_{L^{p/2}}$, where $C_p^* = \frac{(C_{p/2})^2 \epsilon}{1 - C\epsilon} + C_{p/2}$.

Fix $1 \le \alpha \le r$. Let $\tilde{\varphi}^{(\alpha)}$ be the r-tuple of functions $(0, \cdots, 0, 1, 0, \cdots, 0)$ with 1 in the αth position. Let $\varphi^{(\alpha)} = \bar{\partial} \tilde{\varphi}^{(\alpha)} + A_\infty^{(0,1)} \tilde{\varphi}^{(\alpha)}$. Since $\bar{\partial} \tilde{\varphi}^{(\alpha)} = 0$, we have $\|\varphi^{(\alpha)}\|_{L^{p/2}} \le \epsilon \pi^{1/p}$. We can solve $\bar{\partial}_{A_\infty} \tilde{f}^{(\alpha)} = \varphi^{(\alpha)}$ and get $\|\tilde{f}^{(\alpha)}\|_{L_1^{p/2}} \le C_p^* \epsilon \pi^{1/p}$. By Sobolev inequality we have $\|\tilde{f}^{(\alpha)}\|_{L^\infty} \le C_p^\# C_p^* \epsilon \pi^{1/p}$ for some universal constant $C_p^\#$

depending on p. Let $f^{(\alpha)} = \varphi^{(\alpha)} - \tilde{f}^{(\alpha)}$. Then $\bar{\partial}_{A_\infty} f^{(\alpha)} = 0$ and if I denotes the identity $r \times r$ matrix and S denotes the $r \times r$ matrix whose column vectors are $f^{(\alpha)}$ for $1 \leq j \leq r$, then $\|I - S\|_{L^\infty} \leq C_p^{\#} C_p^{*} \epsilon \pi^{1/p}$. Thus for ϵ sufficiently small, the sections $f^{(1)}, \cdots, f^{(r)}$ are $\mathbb{C}$-linearly independent at every point of U. So the holomorphic sections $f^{(1)}, \cdots, f^{(r)}$ form a local basis of E. Therefore E with the complex structure A_∞ is actually a holomorphic vector bundle. From the differential equation $\bar{\partial} f + A_\infty^{(0,1)} f = 0$ and the fact that $A_\infty^{(0,1)}$ has finite L_1^2 norm and f has finite supremum norm it follows that f has finite L_2^2 norm. So when we express the Hermitian metric in terms of this local holomorphic trivialization of E given by $f^{(1)}, \cdots, f^{(r)}$ the Hermitian metric is L_2^2.

(A.7) We now fix a smooth Hermitian metric H_0 of E. Fix a positive number K greater than the L^2 norm of the curvature tensor F_{H_0} of H_0 on M. Consider the set $\mathcal{H}$ of all smooth Hermitian metrics H of E such that $\det H = \det H_0$ and the L^2 norm of F_H on M is no more than K. From the discussion above we can cover M by a finite number of unit coordinate disks U_j so that for each H in $\mathcal{H}$ we have a good gauge for the complex metric connection A_H of H on U_j and the L_1^2 norm on U_j of the matrix-valued 1-form $A_H^{(j)}$ representing A_H is no more that K times some constant depending on M. Moreover, $d^* A_H^{(j)} = 0$ and $A_H^{(j)}$ vanishes at the boundary normals of U_j. Let $\tilde{\mathcal{A}}$ be the set of all connections $\tilde{A}^{(j)}$ on U_j so that $\tilde{A}^{(j)}$ is the weak limit of $A_H^{(j)}$ for H in $\mathcal{H}$. From our discussions above we know that for such connections $\tilde{A}^{(j)}$ we have a Hermitian metric $\tilde{H}$ from the good gauge on U_j which is L_2^2 and we also have the curvature $\tilde{F}$ whose L^2 norm is no more than K.

Among all such connections we minimize the L^2 norm of $\tilde{F}$ and get a minimizing sequence. So we have $A_i^{(j)}$ approaching some $A_\infty^{(j)}$ and have curvature F_∞. The connection $A_\infty = (A_\infty^{(j)})$ is the complex metric connection of some L_2^2 Hermitian metric H_∞ of a holomorphic bundle E_∞.

(A.8) We want to show that there is a nonzero holomorphic homomorphism form E_0 to E_∞. Let $g: E \to E$ be the identity map of E. We use the Hermitian metric H_0 for the domain space E of g and H_i for the image space E of g. To emphasize this fact we write $g: E_0 \to E_i$. Let λ_i be the reciprocal of the L^2 norm of g with respect to the Hermitian metrics H_0 and H_i. Let $g_i: E_0 \to E_i$ be defined by $g_i = \lambda_i g$. Then the L^2 norm of g_i with respect to the Hermitian metrics H_0 and H_i is 1. In terms of the local unitary bases $e_{0\alpha}^{(j)}$ and $e_{i\alpha}^{(j)}$ the matrix valued function $g_i^{(j)}$ defining g_i satisfies the equation

$$0 = \bar{\nabla}_{A_0, A_i} g_i^{(j)} = \bar{\partial} g_i^{(j)} + (A_0^{(j)} - A_i^{(j)})^{(0,1)} g_i^{(j)}$$

on U_j, because the map g_i is a holomorphic section of $\mathrm{End}(E_0, E_i)$, where $\bar{\nabla}_{A_0, A_i}$ means the covariant derivative of $\mathrm{End}(E_0, E_i)$ in the (0,1) direction and $\bar{\partial} g_i^{(j)}$ is the usual (0,1)-derivative of the entries of the matrix $g_i^{(j)}$. Since $(A_0 - A_i)^{(0,1)}$ is bounded in L_1^2 uniformly in i, it follows that for any finite p the L^p norm of $(A_0 - A_i)^{(0,1)} g_i^{(j)}$ is uniformly bounded in i. From the above equation the L_1^p norm of $\bar{\partial} g_i^{(j)}$ on any compact subset of U_j is uniformly bounded in i. By going to a subsequence we can assume that g_i converges strongly in L^2 to some $g_\infty: E_0 \to E_\infty$. It follows that g_∞ is holomorphic and the L^2 norm of g_∞ with respect to the Hermitian metrics H_0 and H_i is 1. In particular g_∞ is not identically zero.

(A.9) We want to show that if $\Lambda F_\infty - \lambda I$ does not vanish identically on M, then the *current* $\Delta(\Lambda F_\infty - \lambda I) = -\bar{\partial}^*_{H_\infty}\bar{\partial}(\Lambda F_\infty - \lambda I)$ is not identically zero. Otherwise $\Delta(\Lambda F_\infty - \lambda' I) = 0$ for any real constant λ' and by integrating it over M against $\Lambda F_\infty - \lambda' I$ we conclude that $\Lambda F_\infty - \lambda' I$ is a holomorphic endomorphism of E_∞. Since $\Lambda F_\infty - \lambda' I$ is Hermitian with respect to H_∞ and its image and kernel, which are holomorphic, must be orthogonal complements of each other with respect to H_∞. So E_∞ splits up into orthogonal holomoprhic summands and ΛF_∞ is equal to a real constant of the identity endomorphism on each of the summands. We claim that the μ-values of all the summands are equal and E_∞ is therefore semi-stable. Otherwise on a coordinate chart we modify the metric on two summands E_1 and E_2 of E_∞ with $\mu(E_1) > \mu(E_\infty)$ and $\mu(E_2) < \mu(E_\infty)$ so that the metric of E_1 as a subbundle of E_∞ (respectively E_2 as a quotient bundle of E_∞) is unchanged. Make the modification so that the global L^2 norm of the second fundamental form B of E_1 is large compared to the global L^2 norm of dB. This is achieved by using a cut-off function $\rho(tz)$ in the modification with t sufficiently small. The global L^2 norm of the trace-free part of the curvature of the new metric is smaller, because the curvature of the new metric of E_1 (respectively E_2) is closer to the curvature of the expected Hermitian-Einstein metric of E_∞. This contradicts the property of the minimizing sequence by considering the semi-universal deformation of E_∞ which has neighboring fibers biholomorphic to E_0. The semi-stability of E_∞ and $\mu(E_0) = \mu(E_\infty)$ imply that g_∞ is a biholomorphism between E_0 and E_∞ and E_∞ is stable and $\Lambda F_\infty = \lambda I$ on E_∞, contradicting the assumption that $\Lambda F_\infty - \lambda I$ is not identically zero.

(A.10) We now show that $\Lambda F_\infty - \lambda I \equiv 0$. Suppose the contrary. Then we can find a smooth $\mathrm{Hom}(E_\infty, E_\infty)$-valued *trace-free* 2-form φ Hermitian with respect to the unitary frame $e^{(j)}_{\infty\alpha}$ so that $\int_M \mathrm{Tr}(\varphi \wedge \Delta(\Lambda F_\infty - \lambda I))$ is strictly negative. Now we want to deform the Hermitian metric in the direction of φ. Let U_t be a smooth family of nonsingular matrices so that the Hermitian matrix $(\det(1 + t\varphi))^{-1/r}(I + t\,\varphi)$ equals $U_t\,\overline{U_t}^{\,t}$, where $\overline{U_t}^{\,t}$ means the complex conjugate transpose of U_t. Now we change the metric H_i (respectively H_∞) to another metric H_{it} (respectively $H_{\infty t}$) so that the unitary frame for H_{it} (respectively $H_{\infty t}$) is obtained from the unitary frame for H_i (repectively H_∞) by applying U_t.

Let $h(t) = (\det(1 + t\varphi))^{-1/r}(I + t\,\varphi)$. Then we have

$$\frac{d}{dt}\,\|F_{H_{\infty t}}\|^2_{L^2} = 2\,\mathrm{Re}\int_M \mathrm{Tr}\left(\frac{\partial h}{\partial t}\,h^{-1}\,\Delta_{H_{\infty t}}F_{H_{\infty t}}\right) = 2\,\mathrm{Re}\int_M \mathrm{Tr}\left(\left(\Delta_{H_{\infty t}}\frac{\partial h}{\partial t}\,h^{-1}\right)F_{H_{\infty t}}\right).$$

We claim that $\Delta_{H_{\infty t}}\frac{\partial h}{\partial t}\,h^{-1}$ and $F_{H_{\infty t}}$ are continuous in t. We have

$$\begin{aligned}
\Delta_{H_{\infty t}}\frac{\partial h}{\partial t}\,h^{-1} &= \Lambda\partial_{H_{\infty t}}\bar\partial_{H_{\infty t}}\frac{\partial h}{\partial t}\,h^{-1} \\
&= \Lambda\partial_{H_{\infty t}}\left(\bar\partial\,\frac{\partial h}{\partial t}\,h^{-1} + \frac{\partial h}{\partial t}\,h^{-1}\,A_{\infty t} - A_{\infty t}\,\frac{\partial h}{\partial t}\,h^{-1}\right) \\
&= \Lambda\partial\left(\bar\partial\,\frac{\partial h}{\partial t}\,h^{-1} + \frac{\partial h}{\partial t}\,h^{-1}\,A_{\infty t} - A_{\infty t}\,\frac{\partial h}{\partial t}\,h^{-1}\right) + \\
&\quad \left(\bar\partial\,\frac{\partial h}{\partial t}\,h^{-1} + \frac{\partial h}{\partial t}\,h^{-1}\,A_{\infty t} - A_{\infty t}\,\frac{\partial h}{\partial t}\,h^{-1}\right)A_{\infty t} - \\
&\quad A_{\infty t}\left(\bar\partial\,\frac{\partial h}{\partial t}\,h^{-1} + \frac{\partial h}{\partial t}\,h^{-1}\,A_{\infty t} - A_{\infty t}\,\frac{\partial h}{\partial t}\,h^{-1}\right).
\end{aligned}$$

Now $A^{(0,1)}_{\infty t} = U_t A^{(0,1)}_\infty U_t^{-1} + \bar\partial U_t\,U_t^{-1}$ (because the $(0,1)$-component of a complex connection, being independent of the metric, obeys the usual transformation rule for connections in frame field change) and $A^{(1,0)}_{\infty t}$ is the negative of the complex conjaguate transpose of $A^{(0,1)}_{\infty t}$. We have

$$\partial\,A^{(0,1)}_{\infty t} = \partial U_t\,A^{(0,1)}_\infty U_t^{-1} + U_t\partial\,A^{(0,1)}_\infty U_t^{-1} - U_t A^{(0,1)}_\infty U_t^{-1}\partial U_t U_t^{-1} + \bar\partial(\bar\partial U_t\,U_t^{-1}).$$

Since $A_\infty^{(0,1)}$ is in L_1^2 and U_t is a smooth function of t and of the variables in M, we conclude that $\partial A_{\infty t}^{(0,1)}$ is actually a smooth function of the variable t with values in the L^2 space of matrix-valued 1-forms on M. Since $A_\infty^{(0,1)}$ is L^4 we conclude that $\Delta_{H_{\infty t}} \frac{\partial h}{\partial t} h^{-1}$ is a smooth function of the variable t with values in the L^2 space of matrix-valued functions on M. Now $F_{\infty t} = F_\infty + \bar{\partial}_{H_\infty}(\partial_{H_\infty} h_t)h_t^{-1}$. We conclude in the same way that $F_{\infty t}$ is a smooth function of the variable t to the L^2 space of matrix-valued functions on M.

The negativity of $\frac{d}{dt} \|F_{H_{\infty t}}\|^2_{L^2}$ at $t = 0$ implies that $\|F_{H_{\infty t}}\|^2_{L^2}$ is strictly decreasing for small values of t. We fix some $t > 0$ so that $\|F_{H_\infty}\|_{L^2} > \|F_{H_{\infty t}}\|_{L^2}$. The connection $A_{\infty t}^{(0,1)} = U_t A_\infty^{(0,1)} U_t^{-1} + \bar{\partial} U_t\, U_t^{-1}$ is the weak limit of $A_{it}^{(0,1)} = U_t A_i^{(0,1)} U_t^{-1} + \bar{\partial} U_t\, U_t^{-1}$ and we have a contradiction. So we conclude that $\Lambda F_\infty - \lambda I$ vanishes on M and the metric H_∞ is Hermitian-Einstein. In particular E with the connection A_∞ is semi-stable.

Since E_0 and E_∞ have the same μ-value and E_0 is stable and E_∞ is semi-stable, we conclude that that holomorphic map g_∞ is a biholomorphism between E_0 and E_∞. Thus through g_∞ the Hermitian metric H_∞ yields a Hermitian-Einstein metric on E_0.

APPENDIX B. Restriction of Stable Bundles

(B.1) Suppose M is a compact complex manifold of complex dimension n with a very ample divisor H. Suppose E is a holomorphic vector bundle over M which is stable (respectively semi-stable) with respect to the polarization H. We would like to show that for m_j $(1 \le j \le n-1)$ sufficiently large the restriction of E to a generic curve C cut out by hypersurfaces in the class H^{m_j} $(1 \leq j \leq n-1)$ is stable (respectively semi-stable). This result is due to Mehta and Ramanathan [M-R1, M-R2]. It clearly implies that the restriction of a stable bundle to a generic hypersurface cut out by a sufficiently high power of an ample divisor is again stable. First we prove the semi-stable case and then explain what modifications are needed for the stable case. We will suppose the contrary and derive a contradiction.

The idea of the proof is to show that if $E|C$ is not semi-stable, then there is a subbundle F of $E|C$ which violates the condition of semi-stability and one tries to extend this subbundle to a subbundle $\tilde{F}$ of E over the whole manifold, which would then violate the semi-stability assumption of E. First we introduce a way of choosing the best subbundle F of $E|C$ that violates the condition of semi-stability.

(B.2) Now let us introduce the notion of such a subbundle. For the time being we change the meaning of the notation E. Let E be a holomorphic vector bundle over a nonsingular curve which is not semi-stable.

A subbundle F of E is called a *strongly contradicting semi-stability* (abbreviated as SCSS) subbundle of E if F is semi-stable and if for every subbundle F' of E properly containing F one has $\mu(F) > \mu(F')$. This notion was introduced in [H-N]. The last condition means that F makes un-semi-stable any subbundle of E that properly contains it.

An equivalent formulation of the last condition is that for any nonzero subbundle Q of E/F one has $\mu(Q) < \mu(F)$. For if one denotes by $\tilde{Q}$ the subbundle of E with $\tilde{Q}/F$ isomorphic to Q, then

$$\frac{\deg(F)}{\operatorname{rank}(F)} = \mu(F) > \mu(\tilde{Q}) = \frac{\deg(Q)}{\operatorname{rank}(\tilde{Q})} = \frac{\deg(F) + \deg(Q)}{\operatorname{rank}(F) + \operatorname{rank}(Q)}$$

which implies that $\dfrac{\deg(F)}{\operatorname{rank}(F)} > \dfrac{\deg(Q)}{\operatorname{rank}(Q)}$ and $\mu(F) > \mu(Q)$.

(B.3) We now prove that for any un-semi-stable vector bundle E over a nonsingular curve there exists a unique SCSS subbundle F.

Let $m = \sup \{\mu(F) \mid 0 \neq F \subset E\}$. Since E is not stable, we have $m > \mu(E)$. Among all the subbundles F with $\mu(F) = m$ we choose one, say F_0, with maximum rank. If $0 \neq F' \subset F_0$ we have $\mu(F') \leq \mu(F_0)$ so that F_0 is semi-stable. On the other hand, if we have a subbundle F' of E strictly containing F_0, then $\operatorname{rank} F' > \operatorname{rank} F_0$ and by the maximality of F_0 we have $\mu(F') < m = \mu(F_0)$. Thus F_0 is SCSS.

Now we want to prove uniqueness. Suppose we have two subbundles F_1, F_2 that are both SCSS. We want to prove that $F_1 = F_2$. Suppose the contrary. Without loss of generality we can assume that F_1 is not contained in F_2. Consider the map $F_1 \to E/F_2$. Let F_1' be its image. The image may not be torsion-free. We factor out the torsion of F_1' and get F_1''. Then we have a monomorphism $F_1'' \to E/F_2$. Since F_1 is semistable, we have $\mu(F_1) \leq \mu(F_1')$. Since F_2 is SCSS, it follows that $\mu(F_1'') < \mu(F_2)$. Since F_1' and F_1'' differ only at a finite number of points and F_1' is contained in F_1'', it follows that $\mu(F_1') \leq \mu(F_1'')$. Hence $\mu(F_1) < \mu(F_2)$. If F_2 is not contained in F_1, by reversing the roles of F_1 and F_2, we get also $\mu(F_2) < \mu(F_1)$, which is a contradiction. Hence F_2 is contained in F_1. By the SCSS property of F_2 we have $\mu(F_2) < \mu(F_1)$, which is again a contradiction. This finishes the proof of the existence and uniqueness of an SCSS subbundle of an unsemi-stable vector bundle over a nonsingular curve.

(B.4) Now we come to the question of patching together the unique SCSS subbundles. Suppose we have a flat family of curves $\pi: X \to S$ and a holomorphic vector bundle E of rank r over X. Assume that X is compact. For every $s \in S$ we denote by $X(s)$ the fiber $\pi^{-1}(s)$ and by $E(s)$ the vector bundle $E|\pi^{-1}(s)$ over $X(s)$.

For each s consider the set $\mathcal{S}(s)$ of all subbundles F of $E(s)$ of rank $< r$ and μ-value $> \mu(E)$ and all torsion-free coherent subsheaves which are limits of such subbundles and then consider the totality $\mathcal{S}$ of all $\mathcal{S}(s)$ for $s \in S$. The set $\mathcal{S}$ has a natural complex structure. We have a projection $\sigma: \mathcal{S} \to S$ whose fibers are $\mathcal{S}(s)$. Let $\tau: \mathcal{S} \times_S X \to S$ be the projection of the fiber product $\mathcal{S} \times_S X$ onto S. We have a torsion-free coherent subsheaf $\mathcal{E}$ of $\tau^* E$ over $\mathcal{S} \times_S X$ so that the subbundle (or torsion-free coherent subsheaf) of $E(s)$ corresponding to the point f of $\mathcal{S}(s)$ is the subbundle (or torsion-free coherent subsheaf) $\mathcal{E}|\{f\} \times X(s)$ of $\tau^* E|\{f\} \times X(s)$.

The reason that we have such a moduli space for subbundles and their limits (which are torsion-free coherent subsheaves) is that we can look at the projectivization $\mathbb{P}(E(s))$ of $E(s)$. Corresponding to a subbundle F of $E(s)$ of rank s we have the projectivization $\mathbb{P}(F)$ of F which is a subvariety of $\mathbb{P}(E(s))$ so that over every point x of $E(s)$ the fiber of $\mathbb{P}(F)$ is an $(s-1)$-dimensional hyperplane in the fiber of $\mathbb{P}(E(s))$ over x. So the moduli space of such subbundles F is a subspace of the moduli space of the deformation of the subvariety $\mathbb{P}(F)$ of $\mathbb{P}(E(s))$ and we can do this with a parameter space S. For compactness we have to argue that the volume of $\mathbb{P}(F)$ is bounded independent of F so that we can apply the theorem of Bishop [Bi] about the convergence of subvarieties of bounded volume.

By tensoring E with the dual of a sufficiently ample line bundle we can assume without loss of generality that E carries a Hermitian metric $H_{\alpha\bar{\beta}}$ with negative curvature tensor $\Omega_{\alpha\bar{\beta}i\bar{j}} = -\partial_i\partial_{\bar{j}}H_{\alpha\bar{\beta}} + H^{\lambda\bar{\mu}}(\partial_i H_{\alpha\bar{\mu}})(\partial_{\bar{j}}H_{\lambda\bar{\beta}})$ with respect to a local holomorphic basis of E. Fix a point P of $X(s)$ and assume that the local holomorphic basis of E is chosen so that at P the first derivative $dH_{\alpha\bar{\beta}}$ vanishes and $H_{\alpha\bar{\beta}}$ equals the Kronecker delta. Let $w^1,\cdots,w^r$ be the holomorphic fiber coordinates of E. Then computation at P using this coordinate system yields

$$\partial\bar{\partial}\log(H_{\alpha\bar{\beta}}\, w^\alpha\, \overline{w^\beta}) = -\frac{\Omega_{\alpha\bar{\beta}}\, w^\alpha \overline{w^\beta}}{h_{\alpha\bar{\beta}}\, w^\alpha \overline{w^\beta}} + \frac{h_{\alpha\bar{\beta}}\, dw^\alpha\, \overline{dw^\beta} - |h_{\alpha\bar{\beta}}\, w^\alpha\, \overline{dw^\beta}|^2}{(h_{\alpha\bar{\beta}}\, w^\alpha\, \overline{w^\beta})^2}$$

which is a positive definite (1,1)-form on $\mathbb{P}(E)$ and we are going to use this positive definite (1,1)-form as a Kähler metric for volume computation. For a subbundle F of $E(s)$ of rank s over $X(s)$ we have to integrate $\left[\partial\bar\partial \log(H_{\alpha\bar\beta}\, w^\alpha \overline{w^\beta})\right]^{s+1}$ over $\mathbb{P}(F)$. Let z be a local holomorphic coordinate of $X(s)$. Since $\Omega_{\alpha\bar\beta i\bar j}$ restricted to $X(s)$ involves $dz\wedge\overline{dz}$, the integration of $\left[\partial\bar\partial \log(H_{\alpha\bar\beta}\, w^\alpha \overline{w^\beta})\right]^{s+1}$ over $\mathbb{P}(F)$ is equal to the integration of

$$(s+1)\left[-\frac{\Omega_{\alpha\bar\beta}\, w^\alpha\overline{w^\beta}}{H_{\alpha\bar\beta}\, w^\alpha\overline{w^\beta}}\right]\left[\frac{H_{\alpha\bar\beta}\, dw^\alpha\, \overline{dw^\beta} - |H_{\alpha\bar\beta}\, w^\alpha\, \overline{dw^\beta}|^2}{(H_{\alpha\bar\beta}\, w^\alpha\, \overline{w^\beta}\,)^2}\right]^s$$

over $\mathbb{P}(F)$. This integral is bounded by a universal constant times the integration of the positive (1,1)-form over $X(s)$ which at every point of $X(s)$ is the supremum of $\Omega_{\alpha\bar\beta i\bar j} w^\alpha\overline{w^\beta}$ over all $(w^1,\cdots,w^r)$ with $H_{\alpha\bar\beta}\, w^\alpha\overline{w^\beta} = 1$. Hence we have the uniform bound for the volumes we want.

Since $\mathcal{S}$ is compact, it has only a finite number of branches $\mathcal{S}_1,\cdots,\mathcal{S}_\ell$. For $1 \le k \le \ell$ let Z_k be the set of points of $\mathcal{S}_k$ where the map $\sigma|\mathcal{S}_k\colon \mathcal{S}_k \to S$ is not locally surjective. Then Z_k is a subvariety of $\mathcal{S}_k$ and $\sigma(Z_k)$ under σ is a subvariety of codimension at least one in S.

Assume that $E(s)$ for a generic point s of S is not semi-stable and for a generic point s in S let $F(s)$ be its unique SCSS subbundle. Choose s_0 in $S - \cup_{k=1}^{\ell}\sigma(Z_k)$ with $E(s_0)$ not semi-stable so that $\mu(F(s_0)) \ge \mu(F(s))$ for all s in $S - \cup_{k=1}^{\ell}\sigma(Z_k)$ with $E(s)$ not semi-stable and $\operatorname{rank}(F(s_0)) \ge \operatorname{rank}(F(s))$ for all s in $S - \cup_{k=1}^{\ell}\sigma(Z_k)$ with $E(s)$ not semi-stable and $\mu(F(s_0)) = \mu(F(s))$. Without loss of generality we can assume that the branch of $\mathcal{S}$ that contains the point f_0 corresponding to the subbundle $F(s_0)$ of $E(s_0)$ is $\mathcal{S}_1$. There exists an open neighborhood U of f_0 in $\mathcal{S}_1$ such that $\sigma(U)$ is an open subset of S. For $f \in U$ the subbundle $\mathcal{E}|\{f\}\times X(\sigma(f))$ has the same μ-value and the rank as those of $\mathcal{E}|\{f_0\}\times X(s_0)$ and hence is the unique SCSS subbundle of $E(\sigma(f))$. Thus σ maps $\mathcal{S}_1$ generically one-to-one onto S. The torsion-free coherent subsheaf $\mathcal{E}|\mathcal{S}_1\times_S X$ of $\tau^*E|\mathcal{S}_1\times_S X$ defines a torsion-free coherent subsheaf of E whose restriction to $X(s)$ for a generic s is the unique SCSS subbundle of $E(s)$. In particular, the determinant bundle of the unique SCSS subbundle of $E(s)$ can be pieced together to form a line bundle over X whose restriction to $X(s)$ is the determinant bundle of the unique SCSS subbundle of $E(s)$.

(B.5) Now we come back to our case of a compact algebraic manifold M and a very ample divisor H of M. Let E be a semi-stable vector bundle of rank ρ over M with respect to the polization H. We continue with our proof that the restriction of E to a complete-intersection generic curve $C_{\mathbf{m}}$ of degree $\mathbf{m} = (m_1, \cdots, m_{n-1})$ is semi-stable for sufficiently large $\mathbf{m}$. Let $\mathbb{P}$ be the product of the complex projective space associated to the vector space $\Gamma(M, H^{m_\upsilon})$ for $1 \leq \mu \leq n-1$. For each $s \in \mathbb{P}$ we have an element s_υ of $\Gamma(M, H^{m_\upsilon})$ determined up to a multiplicative constant. Let Z be the set of all points (x, s) in $M \times \mathbb{P}$ such that x is contained in the zero-set of s_υ for every υ. Let $\sigma\colon Z \to \mathbb{P}$ and $\tau\colon Z \to M$ be induced by the projections of $M \times \mathbb{P}$ onto its two factors.

The collection of the determinant bundles of all the SCSS subbundles for the generic curves $C_{\mathbf{m}}$ give rise to a line bundle $\tilde{L}$ over Z. We want to show that we can construct a line bundle over M so that its restriction to a generic $C_{\mathbf{m}}$ agrees with the restriction of $\tilde{L}$ to $C_{\mathbf{m}}$. We do this one dimension at a time, *i.e.* we take a generic d-dimensional submanifold M_d of M cut out by elements of $\Gamma(M, H^{m_\upsilon})$ for $1 \leq \upsilon \leq n-d$ and find a line bundle on M_2 first and then on M_3, etc. until we get a line bundle on M_d. For notational convenience we give the case of going from $C_{\mathbf{m}}$ to M_2 and so without loss of generality we assume that the complex dimension of M is 2. The argument works the same way when one goes from M_j to M_{j+1}.

This question concerns pushing down a line bundle from the total space of a family of spaces to the base space and it involves showing that the restriction of the line bundle to each fiber is trivial. Let us look at this question of proving the triviality of a line bundle over a general fiber.

First we observe that for any line bundle L over a *reduced irreducible* complex space V to be trivial it is necessary and sufficient that $\Gamma(V,L)$ and $\Gamma(V,L^{-1})$ be nonzero. Suppose we have a flat family $\mathbf{V}$ of reduced irreducible varieties over a parameter complex space S under the map $\pi: \mathbf{V} \to S$ and we have a line bundle L over V such that $L|\pi^{-1}(s)$ is trivial for s in a nonempty open subset of S. Then the zeroth direct images of $\mathcal{O}(L)$ and $\mathcal{O}(L^{-1})$ under π is coherent and of rank at least 1 and by the semicontinuity of the dimension of cohomology group over the fibers of a flat family we conclude that $L|\pi^{-1}(s)$ must be trivial for all s and L must be the pullback of some line bundle over S.

(B.8) Now we apply this observation to the family $\tau: Z \to M$ and a line bundle over Z obtained by modifying the line bundle $\tilde{L}$. The modification of $\tilde{L}$ is needed because in general $\tilde{L}$ is not trivial on most of the fibers of τ. Now $\mathbb{P} = \mathbb{P}_\ell$, where $\dim_{\mathbb{C}}\Gamma(M,H^{m_1}) = \ell + 1$. Since each fiber of τ is $\mathbb{P}_{\ell-1}$, there exists an integer k such that the restriction of $\tilde{L}$ to each fiber of τ is the line bundle over $\mathbb{P}_{\ell-1}$ of Chern class k. Since the projection σ each fiber of τ is projected onto a hyperplane $\mathbb{P}_{\ell-1}$ of $\mathbb{P}$, it follows that the restriction of $\tilde{L}$ to each fiber of τ is the pullback under σ of the line bundle B of $\mathbb{P}$ of Chern class k. Hence $\tilde{L}\otimes\sigma^*B^{-1}$ is trivial on each fiber of τ. We now apply our observation to the line bundle $\tilde{L}\otimes\sigma^{-1}B^*$ and conclude that $\tilde{L}\otimes\sigma^{-1}B^*$ is isomorphic to τ^*A for some line bundle over M. Note that the restriction of $\tilde{L}\otimes\sigma^{-1}B^*$ to any fiber of σ agrees with that of $\tilde{L}$. Hence the restriction of A to a generic $C_{\mathbf{m}}$ agrees with the determinant bundle of the SCSS subbundle of $E|C_{\mathbf{m}}$.

(B.9) We now look at the uniqueness of the extension A. For the uniqueness we have to assume that m_1 is at least three. We claim that the subvariety S of points s of $\mathbb{P}$ where the fiber D_s of σ is not both reduced and irreducible is of codimension at least two in $\mathbb{P}$. The claim will be proved by showing that S is a proper subvariety of an irreducible proper subvariety of $\mathbb{P}$. This irreducible proper subvariety is constructed as follows. Let Y be defined as the subvariety of points of Z where the map $\tau: Z \to \mathbb{P}$ does not have maximum rank. When M is identified with its embedded image in $\mathbb{P}$ by $\Gamma(M,H^{m_1})$, for every $x \in M$ the set $\tau^{-1}(x) \cap Y$ is precisely the set of all hyperplanes of $\mathbb{P}$ that contains a tangent of M. Hence $(\tau|Y): Y \to M$ is a bundle with $\mathbb{P}_{\ell-2}$ as fiber. Hence $\sigma(Y)$ is an irreducible subvariety of codimension ≥ 1 in $\mathbb{P}$ and it is our irreducible proper subvarity of $\mathbb{P}$. It is clear that $\sigma(Y)$ is precisely the set of all points s of $\mathbb{P}$ where the fiber D_s is not regular.

We now show that in the case $m_1 \geq 3$ the subvariety S is contained in $\sigma(Y)$ is not equal to $\sigma(Y)$ and hence of codimension at least two in $\mathbb{P}$. Let $\tilde{M}$ be the embedded image of M in some $\mathbb{P}_N$ by $\Gamma(M,H)$. Fix a point P of $\tilde{M}$. By category arguments one can find a cubic hypersurface of $\mathbb{P}_N$ whose intersection with $\tilde{M}$ is reduced and irreducible and is *singular* at P. This intersection is a fiber D_s of $\sigma: Z \to \mathbb{P}$ with s in $\sigma(Y)$ but not in S.

Suppose A is a line bundle over M such that the restriction of A to a generic D_s for s in some open subset of $\mathbb{P}$ is trivial. Then τ^*A is trivial over $\sigma^{-1}(\mathbb{P} - S)$. Since S is of complex codimension at least two in $\mathbb{P}$, from our in (A.7) it follows that τ^*A is isomorphic to σ^*B for some line bundle B over $\mathbb{P}$. This implies that B is trivial and A is trivial.

(B.10) Now that we have settled the question of piecing together the determinant bundle of the SCSS subbundles, we come back to the extension of the SCSS subbundle itself. We denote by $L_{\mathbf{m}}$ the line bundle over M which extends the determinant subbundle of the SCSS subbundles of $E|C_{\mathbf{m}}$. The next step for the extension of the SCSS subbundle itself is to show that for some subsequence of $\mathbf{m}$ the line bundle $L_{\mathbf{m}}$ can be made independent of $\mathbf{m}$.

We select our subsequence of $\mathbf{m}$ in the following way. Fix positive integers $\alpha_j > 1$ $(1 \le j \le n-1)$. Instead of using $(m_1, \cdots, m_{n-1})$ we are going to use $(\alpha_1^m, \cdots, \alpha_{n-1}^m)$ and denote it by (m). Let $\alpha = \alpha_1 \cdots \alpha_{n-1}$. Similar to $\sigma: Z \to \mathbb{P}$ and $\tau: Z \to M$, for each m we have $\sigma_m: Z_m \to \mathbb{P}_{(m)}$ and $\tau_m: Z_m \to M$. For $\ell > m$ there is some relation between Z_ℓ and Z_m, because the union of $\alpha^{\ell-m}$ fibers of $\sigma_m: Z_m \to \mathbb{P}_{(m)}$ is a fiber of $\sigma_\ell: Z_\ell \to \mathbb{P}_{(\ell)}$. This fiber is a degenerate one. We can join this degenerate fiber to a nonsingular fiber of $\sigma_\ell: Z_\ell \to \mathbb{P}_{(\ell)}$ by a curve in $\mathbb{P}_{(\ell)}$. More precisely we can formulate this observation as follows.

For $\ell > m$ and for any Zariski open subset U_m of $\mathbb{P}_{(m)}$ (respectively U_ℓ of $\mathbb{P}_{(\ell)}$) we can find a point s in $\mathbb{P}_{(\ell)}$ and a nonsingular curve C in $U_\ell \cup \{s\}$ such that $\sigma_\ell^{-1}(C)$ is nonsingular and $\sigma_\ell^{-1}(C) \to C$ is flat and $\sigma_\ell^{-1}(s)$ is a reduced curve with $\alpha^{\ell-m}$ components with at most transversal intersections involving two components at a time and such that each component is a fiber of σ_m over some point of U_m. Here in this precise statement $\sigma_\ell^{-1}(s)$ is the degenerate fiber.

For some nonempty Zariski open subset U_m of $\mathbb{P}_{(m)}$ we can patch the SCSS subbundles F of $E|C_m$ together to get a subbundle F_m of $\tau_m^* E|\sigma^{-1}(U_m)$. We let μ_m be the maximum of such $\mu(F_m|\sigma_m^{-1}(s))$ for $s \in U_m$. By replacing U_m by a smaller nonempty Zariski open subset we can assume that $\mu_m = \mu(F_m|\sigma_m^{-1}(s))$.

From our preceding observation for $\ell > m$ we have a curve C in $\mathbb{P}_{(\ell)}$ with a distinguished point s adapted to U_m and U_ℓ. Then $F_\ell|\sigma_\ell^{-1}(s)$ (after replacing $F_\ell|\sigma_\ell^{-1}(C)$ by its product with a negative power of a local coordinate of C vanishing at s if necessary) consists of subbundles over the $\alpha^{\ell-m}$ branches of $\sigma_\ell^{-1}(s)$ and each of these subbundles has $\mu \leq \mu_m$ by definition of μ_m. Hence $\mu_\ell \leq \alpha^{\ell-m}\mu_m$.

Let d_m be the degree of F_m and let r_m be the rank of F_m. Then by definition $\mu_m = d_m\alpha^m/r_m$. Hence $d_\ell\alpha^\ell/r_\ell \leq \alpha^{\ell-m}d_m\alpha^m/r_m$ and $d_\ell/r_\ell \leq d_m/r_m$. Since r_m must be between 1 and rank E, by choosing a subsequence of m we can assume that d_m and r_m remain constant. This means that in $\mu_\ell \leq \alpha^{\ell-m}\mu_m$ we should have equality and $\det F_\ell|\sigma_\ell^{-1}(s)$ agrees with the union of $\det F_m$ on the $\alpha^{\ell-m}$ branches of $\sigma_\ell^{-1}(s)$. The $\alpha^{\ell-m}$ branches of $\sigma_\ell^{-1}(s)$ are fibers of σ_m over points of U_m. Since we can choose the Zariski open subset U_m to avoid any subvariety of U_m, it follows that L_ℓ and L_m on any generic C_m. By the uniqueness result of (A.9) we conclude that L_ℓ agrees with L_m over M. We denote this common line bundle over M by L.

(B.11) We are going to use Plücker coordinates to extend the SCSS subbundle itself. Let r be the $\mathbb{C}$-rank of the SCSS subbundle F of $E|C_m$ for some generic C_m. We denote C_m simply by C. Let ρ be the $\mathbb{C}$-rank of E.

We take a holomorphic local basis $v_1^{(i)},\cdots,v_r^{(i)}$ of F and consider the Plücker coordinates which are the $\binom{\rho}{r}$ coordinates of $v_1^{(i)}\wedge\cdots\wedge v_r^{(i)}$ with respect to a local basis of E. The subbundle F of $E|C$ is determined by these Plücker coordinates. A more abstract way of seeing this is look at the homomorphism $\Lambda^r F \to \Lambda^r E|C$ induced from the inclusion homomrophism $F \to E|C$. Equivalently one can look at the homomorphism $1 \to (\Lambda^r F)^{-1}(\Lambda^r E|C)$ which is its tensor product with the identity homomorphism of $(\Lambda^r F)^{-1}$. Recall that the determinant bundle $\Lambda^r F$ is the restriction to C of the line bundle L over M which is independent of m when m belongs to a suitable subsequence. The Plücker coordinates correspond to local coordinates of this section in $\Gamma(C, L^{-1} \otimes (\Lambda^r E))$.

We would like to look at the converse problem. Suppose we have a nonzero element of $\Gamma(M, L^{-1} \otimes \Lambda^r E)$. Under what condition can we recover a holomorphic subbundle of E over M of rank r ? The Plücker coordinates of a Grassmannian satisfies a number of quadratic equations. So at a point x of M, we have a complex analytic subset Σ_x of the fiber $(L^{-1} \otimes \Lambda^r E)_x$ consisting of points satisfying those quadratic equations of a Grassmannian. The totally Σ of such complex-analytic subsets Σ_x is a complex-analytic subset of $L \otimes \Lambda^r E$. We can describe Σ by equations as follows.

For some positive integer k we have a finite number of holomorphic sections $t_1, \cdots, t_\ell$ of $\Gamma(M, (L^{-1} \otimes \Lambda^r E)^* \otimes H^k)$ generating $\Gamma(M, (L^{-1} \otimes \Lambda^r E)^* \otimes H^k)$. We have a finite number of quadratic polynomials $P_j(X_1, \cdots, X_\ell)$ such that Σ is the common zero-set of $P_j(t_1, \cdots, t_\ell)$. Take an element φ of $\Gamma(M, L^{-1} \otimes \Lambda^r E)$. Consider the set $\Sigma(\varphi)$ of all points x of M such that $\varphi(x)$ belongs to Σ. Geometrically it means the set of points x such that $\varphi(x)$ is a collection of Plücker coordinates. The complex-analytic subset $\Sigma(\varphi)$ of M is the common zero-set of the elements $P_j(t_1\varphi, \cdots, t_\ell\varphi)$ of $\Gamma(M, H^{2k})$. If $\Sigma(\varphi)$ is the whole manifold M, then φ gives rise to a holomorphic subbundle of E at points where φ is nonzero.

Consider a branch V of $\Sigma(\varphi)$ of dimension $d < n$. Then the volume of V with respect to the Kähler metric defined by H is bounded by $(2k)^{n-d} H^n$. Here H^n denotes the intersection number of n copies of H. We claim that for m sufficiently large V cannot contain a generic curve C_m. Now C_m is the intersection of the zero-sets Z_j of $f_j \in \Gamma(M, H^{m_j})$, $1 \leq j \leq n-1$. For a generic C_m the intersection V with $Z_1, \cdots, Z_j$ has dimension $d - j$. The area of the intersection of V with $Z_1, \cdots, Z_{n-d-1}$ is no more than $(2k)^{n-d} m_1 \cdots m_{n-d-1}$ which is less than the volume $m_1 \cdots m_{n-1}$ when m is sufficiently large. Thus unless $\Sigma(\varphi)$ is all of M, the set $\Sigma(\varphi)$ cannot contain a generic C_m for m sufficiently large. Thus we have the following important conclusion.

There exists m_0 such that for $m > m_0$ and for an element φ of $\Gamma(M, L^{-1} \otimes \Lambda^r E)$, if the restriction $\varphi | C_m$ to C_m comes from some holomorphic vector subbundle F of $E | C_m$, then we can construct a reflexive coherent subsheaf of E over M from φ which extends F.

Since H is ample, by the vanishing theorem for ample line bundles we conclude that for m sufficiently large the restriction map $\Gamma(M, L^{-1} \otimes \Lambda^r E) \to \Gamma(C_m, L^{-1} \otimes \Lambda^r E)$ is surjective. So from the SCSS subbundle F of E over C_m we obtain an element of $\Gamma(C_m, L^{-1} \otimes \Lambda^r E)$ which can be extended to an element φ of $\Gamma(M, L^{-1} \otimes \Lambda^r E)$. When $m > m_0$ this element φ of $\Gamma(M, L^{-1} \otimes \Lambda^r E)$ gives us an extension of the subbundle F of $E | C_m$ to a subbundle $\tilde{F}$ of E over M. This subbundle $\tilde{F}$ yields a contradiction to the semistability of E because $\mu(\tilde{F}) = \mu(F)$ and $\mu(E) = \mu(E | C_m)$.

(B.12) We now look at the case of the restriction of stable bundles to curves. For the stable case we have to replace the SCSS subbundle by something else in our argument. For the time being we use the temporary notation that E is a semi-stable holomorphic vector bundle over a nonsingular curve C which is not stable. Since E is not stable, we have a proper subbundle F of E with $\mu(F) = \mu(E)$. Among all such proper subbundles of E choose one with smallest rank and call it F_1. Then F_1 is stable. The quotient E/F_1 is semi-stable. If it is not stable, we can choose a proper subbundle F_2 of E such that F_2/F_1 is the subbundle in E/F_1 of the smallest rank such that $\mu(F_2/F_1) = \mu(E/F_1)$. We can repeat the argument with E/F_1 replaced by E/F_2 and do this inductively. So we come up with $0 = F_0 \subset F_1 \subset F_2 \subset \cdots \subset F_p = E$ such that F_i/F_{i-1} is stable for $1 \leq i \leq p$. We call this sequence of nested subbundles a *stable filtration* of E. It was introduced by Seshadri. A stable filtration is not unique, but the graded bundle $\oplus_{i=1}^p F_i/F_{i-1}$ is unique. The reason for the uniqueness of the graded module is the following. Consider the category $\mathcal{C}$ of all semi-stable bundles F over M with $\mu(F) = \mu(E)$ and bundle-homomorphisms. The kernel, the image, and the cokernel of any bundle homomorphism $F' \to F''$ between objects F' and F'' in $\mathcal{C}$ also belong to $\mathcal{C}$, because both F'' and F'' are semi-stable with the same μ-value. Moreover, the direct sum of two objects in $\mathcal{C}$ is also an object in $\mathcal{C}$. So $\mathcal{C}$ is a so-called abelian category. An object F of $\mathcal{C}$ is called simple if any bundle-homomorphism from F to an object of $\mathcal{C}$ is either zero or a monomorphism. It is clear from the definition of a stable bundle that an object F of $\mathcal{C}$ is simple if and only if F is a stable bundle. An increasing sequence of subobjects $0 = F_0 \subset F_1 \subset F_2 \subset \cdots \subset F_p = F$ of an object F in $\mathcal{C}$ is called a Jordan-Hölder series of F if each F_i/F_{i-1} is a simple object in $\mathcal{C}$ for $1 \leq i \leq p$. The object $\oplus_{i=1}^p F_i/F_{i-1}$ of $\mathcal{C}$ is called the graded object of the Jordan-Hölder series of F. By the standard simple arguement in an abelian category we know that the graded objects of two Jordan-Hölder series of an object in $\mathcal{C}$ must be isomorphic. A stable filtration of E is simply a Jordan-Hölder series of E as an object in $\mathcal{C}$. So we have the uniqueness of the graded bundle of a stable filtration. Let $r_j = \operatorname{rk} F_j$ and $L^{(j)} = \det (F_j/F_{j-1})$. Then r_j and $L^{(j)}$ are unique. The notion of stable filtration was introduced by Seshadri [Se].

As in (B.10) we let $\Sigma^{(r)}$ be the complex analytic subset of $\Lambda^r E$ consisting of decomposable elements of $\Lambda^r E$ (*i.e.* elements which are exterior products of precisely r elements of E). In other words $\Sigma^{(r)}$ is the set of points satisfying the defining quadratic equations of a Grassmannian. When $1 \leq r < \operatorname{rk} E$ and L is a line bundle over C with $\mu(L) = \mu(E)$, the pair (r, L) is equal to some $(r_j, L^{(j)})$ if and only if there exists a nonzero homomorphism f from L to $\Lambda^r E$ whose image is contained in $\Sigma^{(r)}$. The "only if" part is clear, because we can choose f to be the homomorphism obtained by taking the r_j-fold exterior product of the inclusion map $F_j \to E$. Conversely, we let $\tilde{L}$ be the subbundle of $\Lambda^r E$ generated by the image of L. Since $\tilde{L}$ is contained in $\Sigma^{(j)}$ it defines a subbundle F of E of rank r whose determinant bundle is $\tilde{L}$. Since $f(L) \subset \tilde{L}$, it follows that $\mu(L) \leq \mu(\tilde{L})$. Since E is semi-stable, $\mu(F) \leq \mu(E)$. It follows from $\mu(L) = \mu(E)$ that $\mu(\tilde{L}) = \mu(F) = \mu(E) = \mu(L)$. Hence L is isomorphic to $\tilde{L}$ under f and is the determinant bundle of F. We can now start with a stable filtration of the semi-stable bundle F and complete it to a stable filtration of E. So (r,L) is equal to some $(r_j, F^{(j)})$ from a stable filtration of E.

We now use the notations of (B.4). We assume that $E(s)$ is semi-stable over $X(s)$ for a generic $s \in S$. We consider the set $\mathcal{S}(s)$ of all proper subbundles F of $E(s)$ whose μ-value equals $\mu(E(s))$ and carry out the construction as in (B.4) and get $\mathcal{S}_1, \cdots, \mathcal{S}_\ell$. We throw away those $\mathcal{S}_j$ with the rank as some other $\mathcal{S}_k$ and we get a subset of $\mathcal{S}_1, \cdots, \mathcal{S}_\ell$. We denote this subset again by $\mathcal{S}_1, \cdots, \mathcal{S}_\ell$. Let r_j be the rank of $\mathcal{S}_j$. For $f \in \mathcal{S}_j$ with $\sigma(f)$ outside some suitable subvariety of codimension ≥ 1 in S, the determinant bundle $L^{(j)}(\sigma(f))$ of the subbundle $\mathcal{E}|\{f\} \times X(\sigma(f))$ of $E(\sigma(f))$ depends only on $\sigma(f)$ and the pair $(r_j, L^{(j)}(\sigma(f)))$ comes from a stable filtration of the semi-stable bundle $E(\sigma(f))$. So applying the argument of (B.4) to $L^{(j)}(s)$ instead of the unique SCSS subbundle of $E(s)$, we obtain a line bundle $L^{(j)}$ over X whose restriction to $X(s)$ is $L^{(j)}(s)$ for a generic s.

So in the notations of (B.10), instead of one line bundle L_m over M whose restriction to a generic C_m is the determinant bundle of the SCSS subundle of $E|C_m$, we have line bundles $L_m^{(j)}$ over M whose restriction to a generic C_m is the determinant bundle of the subbundle F_j in a stable filtration $F_0 \subset F_1 \subset F_2 \subset \cdots \subset F_r$ of $E|C_m$. As in (A.9) we have to know how $L_m^{(j)}$ changes with m.

We use the notations of (B.10). For $\mathbf{m} = (\alpha_1^m, \cdots, \alpha_{n-1}^m)$ we denote $L_{\mathbf{m}}^{(j)}$ by $L_m^{(j)}$ and denote $C_{\mathbf{m}}$ as C_m. For $\ell > m$ we again use the technique of degenerate fibers as in (B.10). We can choose a 1-parameter family of C_ℓ parametrized by a curve C nonsingular at s so that the curve C_ℓ corresponding to s is a union of $\alpha^{\ell-m}$ curves C_m with at most transversal intersections involving two curves at a time. We can do this so that for prescribed Zariski open subsets U_m of $\mathbb{P}_{(m)}$ and U_ℓ of $\mathbb{P}_{(\ell)}$ the curve $C - \{s\}$ is in U_ℓ and s is in U_m. We have a nonzero homomorphism from $L_\ell^{(j)}$ to $\Lambda^{r_j} E|C_\ell$ so that the image is inside the set of decomposable elements. We also have such a nonzero homomorphism over the curve C_ℓ corresponding to the point s. After checking the relevant μ-values, we conclude that $L_\ell^{(j)}$ equals some $L_m^{(k)}$. Thus there exists some j such that $L_m^{(j)}$ is independent of m for m sufficiently large. We can now repeat the argument in (B.11) of extending subbundles. This concludes the proof of the stable case.

CHAPTER 2. KÄHLER-EINSTEIN METRICS FOR THE CASE OF NEGATIVE AND ZERO ANTICANONICAL CLASS

§1. *Monge-Ampere Equation and Uniqueness.*

(1.1) Suppose M is a compact Kähler manifold of complex dimension m. We would like to prove in this chapter the existence of a Kähler-Einstein metric on M when the anticanonical class of M is either negative or zero. A Kähler-Einstein metric means a Kähler metric whose Ricci curvature is a constant multiple of the Kähler metric. First we formulate the problem in the form a Monge-Ampère equation.

When the anticanonical class is negative, we assume that the given Kähler form is in the canonical class. We use c to denote -1 when the anticanonical class is negative and use c to denote 0 when the anticanonical class is zero. Let $g_{i\bar{j}}$ be the Kähler metric of M and $R_{i\bar{j}}$ be the Ricci curvature expressed in terms of local coordinates $z^1, \cdots, z^m$ of M. The Ricci curvature $R_{i\bar{j}}$ is given by the formula $R_{i\bar{j}} = -\partial_i\partial_{\bar{j}} \log \det(g_{k\bar{\ell}})$, where ∂_i means $\dfrac{\partial}{\partial z^i}$ and $\partial_{\bar{j}}$ is the complex conjugate of ∂_j. Since $R_{i\bar{j}}$ and $cg_{i\bar{j}}$ are in the class, $R_{i\bar{j}} + cg_{i\bar{j}} = \partial_i\partial_{\bar{j}}F$ for some smooth function F on M. We are going to prove the existence of a Kähler-Einstein metric in the same class as $g_{i\bar{j}}$. Any Kähler metric $g'_{i\bar{j}}$ in the same class as $g_{i\bar{j}}$ is of the form $g_{i\bar{j}} + \partial_i\partial_{\bar{j}}\varphi$ for some smooth function φ on M. Consider the Monge-Ampère equation

$$\frac{\det(g_{i\bar{j}} + \varphi_{i\bar{j}})}{\det(g_{i\bar{j}})} = e^{-c\varphi + F}. \tag{1.1.1}$$

The solution of this Monge-Ampère equation with $g_{i\bar{j}} + \partial_i\partial_{\bar{j}}\varphi$ positive definite everywhere would give us a Kähler-Einstein metric for the following reason. Taking $\partial_i\partial_{\bar{j}}$ of the logarithm of both sides of the Monge-Ampère equation yields

$$\begin{aligned} -R'_{i\bar{j}} + R_{i\bar{j}} &= -c\partial_i\partial_{\bar{j}}\varphi + \partial_i\partial_{\bar{j}}F \\ &= -c(g'_{i\bar{j}} - g_{i\bar{j}}) + (R_{i\bar{j}} - cg_{i\bar{j}}) \\ &= -cg'_{i\bar{j}} + R_{i\bar{j}} \end{aligned}$$

from which it follows that $R'_{i\bar{j}} = c\, g'_{i\bar{j}}$, where $R'_{i\bar{j}}$ is the Ricci curvature of $g'_{i\bar{j}}$. Take $1 > \epsilon > 0$ and we will specify ϵ later. We are going to solve the Monge-Ampère equation by the continuity method. More precisely we introduce a parameter $t \in [0,1]$ into the equation and consider the following equation with parameter

$$\frac{\det(g_{i\bar{j}} + \varphi_{i\bar{j}})}{\det(g_{i\bar{j}})} = e^{-c\varphi + tF} \tag{1.1.2}$$

in the case $c = -1$ and the following equation with parameter

$$\frac{\det(g_{i\bar{j}} + \varphi_{i\bar{j}})}{\det(g_{i\bar{j}})} = A_t e^{tF} \tag{1.1.3}$$

in the case $c = 0$, where $A_t = (\mathrm{Vol}\, M)\left[\int_M e^{tF}\right]^{-1}$ and the integration is with respect to the volume form of $g_{i\bar{j}}$. The constant A_t is inserted to make the integral of $A_t e^{tF}$ over M equal $\mathrm{Vol}\, M$ for the sake of compatibility, because the integral of $\frac{\det(g_{i\bar{j}} + \varphi_{i\bar{j}})}{\det(g_{i\bar{j}})}$ over M with respect to the volume form of $g_{i\bar{j}}$ is easily seen to be equal to the volume of M when one uses the language of exterior product and differential forms instead of the language of determinants.

For notational convenience, to unify the two equations (1.1.2) and (1.1.3) we use the convention that $A_t = 1$ when $c = -1$. With this convention the two equations (1.1.2) and (1.1.3) are both special cases of the following equation

$$(1.1.4) \qquad \frac{\det(g_{i\bar{j}} + \varphi_{i\bar{j}})}{\det(g_{i\bar{j}})} = A_t e^{-c\varphi + tF} .$$

Let k be an integer ≥ 3.

(1.2) Consider first the case $c = -1$. Let T be the set of all $t \in [0,1]$ for which the equation (1.1.2) admits a solution φ in $C^{k+\epsilon}$ with $g_{i\bar{j}} + \partial_i\partial_{\bar{j}}\varphi$ positive definite everywhere. The equation (1.1.1) will be solved if we can prove that T is both an open subset and a closed subset of the connected set $[0,1]$, because obviously T contains 0 (with the solution $\varphi = 0$ for (1.1.2)) and the equation (1.1.1) is simply the equation (1.1.2) with $t = 1$. Here $C^{k+\epsilon}$ means that the space of functions whose derivatives of order $\leq k$ have finite Hölder norm of exponent ϵ.

Openness of the set T is proved by using the inverse function theorem. Suppose we have a solution φ_{t_0} in $C^{k+\epsilon}$ which is a solution of the equation (1.1.2) for $t = t_0$ with $g_{i\bar{j}} + \partial_i\partial_{\bar{j}}\varphi_{t_0}$ positive definite everywhere. First consider the case $c = -1$. Consider the operator Ψ mapping φ in $C^{k+\epsilon}$ near φ_{t_0} to $\log \dfrac{\det(g_{i\bar{j}} + \varphi_{i\bar{j}})}{\det(g_{i\bar{j}})} + c\varphi$ in $C^{k-2+\epsilon}$. Then the differential $d\Psi$ of Ψ at φ_{t_0} evaluated at the direction $\dot{\varphi}$ is $\Delta_{\varphi_{t_0}}\dot{\varphi} + c\dot{\varphi}$, where $\Delta_{\varphi_{t_0}}$ is the Laplacian with respect to the Kähler metric $g_{i\bar{j}} + \partial_i\partial_{\bar{j}}\varphi_{t_0}$. (For the computation of $d\Psi$ one considers a family of φ depending on a real parameter s so that φ is equal to φ_{t_0} at $s = 0$ and then differentiates $\Psi(\varphi) = \log \dfrac{\det(g_{i\bar{j}} + \varphi_{i\bar{j}})}{\det(g_{i\bar{j}})} + c\varphi$ with respect to s

and set $s = 0$ and $\dot{\varphi} = \frac{\partial\varphi}{\partial s}\Big|_{s=0}$.) The operator $\dot{\varphi} \to (\Delta_{\varphi_{t_0}} + c)\dot{\varphi}$ is bijective from the Banach space $C^{k+\epsilon}$ onto the Banach space $C^{k-2+\epsilon}$. So $d\Psi$ at φ_{t_0} is an isomorphsim and we have a solution φ_t of the equation $\Psi(\varphi_t) = tF$ for t sufficiently close to t_0. Thus we have the openness of T at t_0.

(1.3) Now consider the case $c = 0$. Let B_1 be the set of all elements of $C^{k+\epsilon}$ whose integral over M with respect to the volume form of $g_{i\bar{j}}$ vanishes. Let B_2 be the set of all elements of $C^{k-2+\epsilon}$ whose integral over M with respect to the volume form of $g_{i\bar{j}}$ equals the volume of M. Let T be the set of all $t \in [0,1]$ for which the equation (1.1.3) admits a solution φ in B_1 with $g_{i\bar{j}} + \partial_i\partial_{\bar{j}}\varphi$ positive definite everywhere. Again the equation (1.1.1) will be solved if we can prove that T is both an open and closed subset of the connected set $[0,1]$. To prove the openness of T by the inverse function theorem, we assume that $\varphi_{t_0} \in B_1$ is a solution of the equation (1.1.3) for $t = t_0$ with $g_{i\bar{j}} + \partial_i\partial_{\bar{j}}\varphi_{t_0}$ positive definite everywhere. Consider the operator Ψ mapping an element φ in B_1 which is near φ_{t_0} to $\frac{\det(g_{i\bar{j}} + \varphi_{i\bar{j}})}{\det(g_{i\bar{j}})}$ which is in B_2. The differential $d\Psi$ of Ψ at φ_{t_0} evaluated at the direction $\dot{\varphi}$ is $\Delta_{\varphi_{t_0}}\dot{\varphi}$ and the tangent space of B_2 consists of all $C^{k-2+\epsilon}$ functions whose integrals over M with respect to $g_{i\bar{j}}$ vanishes. Hence $d\Psi$ is invertible and we have the openness of the set T at $t = t_0$.

(1.4) In both cases, to show that the set T is closed, we need *a priori* estimates. More precisely, we have to show that if $\varphi \in C^{k+\epsilon}$ satisfies the equation (1.1.4) with $g_{i\bar{j}} + \partial_i\partial_{\bar{j}}\varphi$ positive definite everywhere, then the $C^{k+\epsilon}$ norm of φ is bounded by a constant independent of t and φ but dependent on k. We observe that since φ satisfies the equation, by interior elliptic Schauder estimates [G-T, p.90, Th.6.2] we can get *a priori* $C^{\nu+\epsilon}$ estimates of φ in terms of the $C^{2+\epsilon}$ bound of φ for $\nu \geq 2$. The reason is as follows. Assume that we have *a priori* $C^{2+\epsilon}$ estimates for φ. Locally we write $g_{i\bar{j}} = \partial_i\partial_{\bar{j}}\psi$ for some smooth local function ψ. By taking ∂_ℓ log of both sides of the equation (1.1.4), we get

$$\Delta_\varphi(\partial_\ell(\psi+\varphi)) = \Delta(\partial_\ell\psi) - c\partial_\ell\varphi + t\partial_\ell F, \tag{1.4.1}$$

where Δ_φ is the (negative) Laplace operator with respect to $g_{i\bar{j}} + \partial_i\partial_{\bar{j}}\varphi$ and Δ is the (negative) Laplace operator with respect to $g_{i\bar{j}}$. Since from the equation (1.1.4) we have an *a priori* positive lower bound on the determinant of $g_{i\bar{j}} + \partial_i\partial_{\bar{j}}\varphi$, it follows from the *a priori* C^2 bound of φ that we have an *a priori* positive lower bound for the smallest eigenvalue of the Hermitian matrix $g_{i\bar{j}} + \partial_i\partial_{\bar{j}}\varphi$. Inductively for $0 \leq \nu < \infty$, by interior Schauder estimates [G-T, p.90, Th.6.2] for linear elliptic equations with coefficients in $C^{\nu+\epsilon}$, from equation (1.4.1) we obtain the *a priori* $C^{\nu+3+\epsilon}$ norm estimate of φ in terms of the $C^{\nu+1+\epsilon}$ norm of φ. This actually shows that the solution φ in $C^{k+\epsilon}$ must be infinitely differentiable. So we need only get *a priori* $C^{2+\epsilon}$ estimates for φ. We break this up into three steps: (i) the zeroth order estimate, (ii) the second order estimate, and (iii) the Hölder estimate for the second derivative. Step (iii) depends on step (ii) which in turn depends on step (i).

(1.5) Before we prove the three *a priori* estimates, we would like to discuss the uniqueness of Kähler-Einstein metrics. Suppose we have a Kähler-Einstein metric $g'_{i\bar{j}}$ in the same class as $g_{i\bar{j}}$. Then we can write $g'_{i\bar{j}} = g_{i\bar{j}} + \partial_i\partial_{\bar{j}}\varphi$ for some smooth function φ. The Kähler-Einstein condition of $g_{i\bar{j}}$ means that $R'_{i\bar{j}} = cg'_{i\bar{j}}$. That is,

$$-\partial_k\partial_{\bar{\ell}} \log\det(g_{i\bar{j}} + \partial_i\partial_{\bar{j}}\varphi) = c(g_{k\bar{\ell}} + \partial_k\partial_{\bar{\ell}}\varphi). \tag{1.5.1}$$

Since we assume that $g_{i\bar{j}}$ is in the canonical class when the anticanonical class is negative, we have a smooth function F such that $R_{i\bar{j}} - cg_{i\bar{j}} = \partial_i\partial_{\bar{j}}F$. So we can rewrite (1.5.1) as

$$-\partial_k\partial_{\bar{\ell}} \log \det (g_{i\bar{j}} + \partial_i\partial_{\bar{j}}\varphi) = R_{k\bar{\ell}} - \partial_k\partial_{\bar{\ell}}F + c\partial_k\partial_{\bar{\ell}}\varphi$$
$$= -\partial_k\partial_{\bar{\ell}} \log \det(g_{i\bar{j}}) - \partial_k\partial_{\bar{\ell}}F + c\partial_k\partial_{\bar{\ell}}\varphi.$$

After we add a suitable constant to F, we get equation (1.1.1). Uniqueness of the Kähler-Einstein metric in the given class $g_{i\bar{j}}$ is equivalent to the uniqueness of the solution of (1.1.1) with $g_{i\bar{j}} + \partial_i\partial_{\bar{j}}\varphi_{t_0}$ positive definite everywhere (and in the case $c = 0$ with the additional assumption that the integral of φ over M with respect to $g_{i\bar{j}}$ vanishes).

Consider first the case $c = -1$. Suppose we have two solutions φ and φ of (0.1). Let $g'_{i\bar{j}} = g_{i\bar{j}} + \partial_i\partial_{\bar{j}}\varphi$ and $\theta = \psi - \varphi$. We divide the equation for ψ by the equation for φ and get

$$\frac{\det(g'_{i\bar{j}} + \theta_{i\bar{j}})}{\det(g'_{i\bar{j}})} = e^{\theta}. \tag{1.5.2}$$

Let P be the point where the supremum of θ is achieved. Then since the matrix $(\theta_{i\bar{j}})$ is is negative semidefinite and the matrix $(g'_{i\bar{j}} + \theta_{i\bar{j}})$ is $\leq$ the matrix $(g'_{i\bar{j}})$. So the determinant of $(g'_{i\bar{j}} + \theta_{i\bar{j}})$ is $\leq$ the determinant of $(g'_{i\bar{j}})$. Hence the left-hand side of (1.5.2) is ≤ 1 and e^{θ} is ≤ 1 and $\theta \leq 0$ at P. This means that θ is ≤ 0 on M. In the same way by considering the point where the infimum of θ is achieved, one concludes that θ is ≥ 0 on M. Hence θ is identically zero and $\varphi = \psi$.

Now consider the case $c = 0$. Again suppose we have two solutions φ and φ of (1.1.1). Let $g'_{i\bar{j}} = g_{i\bar{j}} + \partial_i\partial_{\bar{j}}\varphi$ and $g''_{i\bar{j}} = g_{i\bar{j}} + \partial_i\partial_{\bar{j}}\psi$ and $\theta = \psi - \varphi$. Let $\omega' = \sqrt{-1}g'_{i\bar{j}}\, dz^i \wedge \overline{dz^j}$ and $\omega'' = \sqrt{-1}g''_{i\bar{j}}\, dz^i \wedge \overline{dz^j}$. From the equation (1.1.1) for φ and for ψ we get $\omega'^m - \omega''^m = 0$ which can be rewritten as

$$(1.5.3) \qquad \sqrt{-1}\,\partial\bar\partial\theta \wedge \left[\Sigma_{\nu=0}^{m-1} \omega'^{\nu}\omega''^{m-1-\nu}\right] = 0.$$

Since both ω' and ω'' are positive definite (1,1)-forms, the equation (1.5.3) is a linear elliptic equation in θ if we consider contributions from $\Sigma_{\nu=0}^{m-1} \omega'^{\nu}\omega''^{m-1-\nu}$ as known variable coefficients. Since the equation has no zero-order term and the maximum of θ is achieved at an interior point, by the strong maximum principle of E. Hopf the function θ must be constant. This concludes the proof of the uniqueness of Kähler-Einstein metrics. Let us now summarize the result which will be proved after the required *a priori* estimates are established later.

Theorem. Suppose M is a compact Kähler manifold whose anticanonical line bundle is either negative or trivial. Assume that the given Kähler class is in the canonical class when the anticanonical line bundle is negative. Then there exists a unique Kähler-Einstein metric in the given Kähler class.

The proof of the theorem for the case of trivial canonical line bundle also yields the result that on a compact Kähler manifold any real (1.1)-form representing the anticanonical class is the Ricci curvature form of some Kähler metric in the same class as the original Kähler metric.

§2. *Zeroth order estimates.*

(2.1) The zeroth order estimate for the case of negative first Chern class is very easy and the argument is the same as the uniqueness of the solution of the equation (0.1) for $c = -1$. Since we have the Monge-Ampère equation

$$\frac{\det(g_{i\bar j} + \varphi_{i\bar j})}{\det(g_{i\bar j})} = e^{\varphi + tF}.$$

At the point where φ achieves its maximum, the matrix $(\varphi_{i\bar j})$ is negative semidefinite and the matrix $(g_{i\bar{j}} + \varphi_{i\bar{j}})$ is $\leq$ the matrix $(g_{i\bar{j}})$. So the determinant of $(g_{i\bar{j}} + \varphi_{i\bar{j}})$ is $\leq$ the determinant of $(g_{i\bar{j}})$. Hence $e^{\varphi + tF}$ is ≤ 1 and $\sup_M \varphi \leq \sup_M(-tF)$. In the same way one concludes that $-\inf_M \varphi \leq \sup_M(tF)$. So we have the zeroth order *a priori* estimate for φ.

(2.2) The zeroth order estimate is more complicated for the case of zero first Chern class. It is done by using the Moser iteration technique [Mo2]. The idea of the Moser iteration technique for a linear ellitpic equation is as follows. One multiplies the linear elliptic equation by a power of the solution and integrates by parts to get an estimate of the L^2 norm of the derivative of a power of the solution in terms of its L^2 norm of the same power of the solution. Then one uses the Sobelev lemma to get estimates of L^p norm of the solution for large p in terms of its L^2 norm and finally get the supremum norm estimates in the limiting case of $p \to \infty$. Here we have a nonlinear elliptic equation instead of a linear one, but still we can imitate this procedure of integration by parts and using Sobolev lemma to estimate L^p norm for large p in terms of L^2. The only difference is that one has to additional terms from the process of integration by parts because of the nonlinearity. These terms are essentially harmless so far as the required inequality is concerned because of the positivity of the matrix $(g_{i\bar{j}} + \varphi_{i\bar{j}})$.

(2.3) Let ω be the Kähler form of the original metric $g_{i\bar{j}}$. Consider an increasing function $h(\varphi)$ of φ, which later will be a function corresponding to a power of φ. Now we do the argument corresponding to the process in the Moser iteration technique [Mo2] of multiplying the linear elliptic equation by a power of the solution and integrating by parts.

$$\int(\omega + \sqrt{-1}\partial\bar{\partial}\varphi)^m h(\varphi) = \int(\Sigma_{\nu=0}^m \binom{m}{\nu}\omega^{m-\nu}(\sqrt{-1}\partial\bar{\partial}\varphi)^\nu)h(\varphi).$$

Apply Stokes' theorem to

$$\begin{aligned} d(h(\varphi)\omega^{m-\nu}\sqrt{-1}\partial\varphi(\sqrt{-1}\partial\bar{\partial}\varphi)^{\nu-1}) &= h'(\varphi)\omega^{m-\nu}\sqrt{-1}\partial\varphi\wedge\bar{\partial}\varphi\wedge(\sqrt{-1}\partial\bar{\partial}\varphi)^{\nu-1} \\ &\quad + h(\varphi)\omega^{m-\nu}(\sqrt{-1}\partial\bar{\partial}\varphi)^\nu. \end{aligned}$$

We get

$$\int h(\varphi)\omega^{m-\nu}(\sqrt{-1}\partial\bar\partial\varphi)^{\nu} = -\int h'(\varphi)\omega^{m-\nu}\sqrt{-1}\partial\varphi\wedge\bar\partial\varphi\wedge(\sqrt{-1}\partial\bar\partial\varphi)^{\nu-1}.$$

Fix a point and choose a local coordinate system such that both $(\partial_i\partial_{\bar j}\varphi)$ and ω are diagonal. Let $\eta = \Sigma_i |\partial_i\varphi|^2\sqrt{-1}dz^i\wedge d\overline{z^i}$. Then

$$\omega^{m-\nu}\wedge\sqrt{-1}\partial\varphi\wedge\bar\partial\varphi\wedge(\sqrt{-1}\partial\bar\partial\varphi)^{\nu-1} = \omega^{m-\nu}\wedge\eta\wedge(\sqrt{-1}\partial\bar\partial\varphi)^{\nu-1}.$$

So far as the coefficient of $|\partial_i\varphi|^2$ is concerned, it suffices to consider

$$\omega^{m-\nu}\wedge\sqrt{-1}dz^i\wedge d\overline{z^i}\wedge(\sqrt{-1}\partial\bar\partial\varphi)^{\nu-1}.$$

Thus the coefficient of $|\partial_i\varphi|^2$ in the integrand of the right-hand side of

$$\begin{aligned}\int\left[(\omega + \partial\bar\partial\varphi)^m - \omega^m\right]h(\varphi) &= \int\left(\Sigma_{\nu=1}^m \binom{m}{\nu}\omega^{m-\nu}(\sqrt{-1}\partial\bar\partial\varphi)^{\nu}\right)h(\varphi)\\ &= -\int h'(\varphi)\Sigma_{\nu=1}^m \binom{m}{\nu}\omega^{m-\nu}\wedge\sqrt{-1}\partial\varphi\wedge\bar\partial\varphi\wedge(\sqrt{-1}\partial\bar\partial\varphi)^{\nu-1}\end{aligned}$$

is

$$\begin{aligned}&h'(\varphi)\Sigma_{\nu=1}^m \binom{m}{\nu}\omega^{m-\nu}\wedge\sqrt{-1}dz^i\wedge d\overline{z^i}\wedge(\sqrt{-1}\partial\bar\partial\varphi)^{\nu-1}\\ &= h'(\varphi)\Sigma_{\nu=1}^m\, m\,\frac{1}{\nu}\binom{m-1}{\nu-1}\omega^{m-\nu}\wedge\sqrt{-1}dz^i\wedge d\overline{z^i}\wedge(\sqrt{-1}\partial\bar\partial\varphi)^{\nu-1}\\ &= h'(\varphi)\Sigma_{\nu=0}^{m-1}\, m\left[\int_{t=0}^1 t^{\nu}dt\right]\binom{m-1}{\nu}\omega^{m-\nu}\wedge\sqrt{-1}dz^i\wedge d\overline{z^i}\wedge(\sqrt{-1}\partial\bar\partial\varphi)^{\nu-1}\\ &= m\, h'(\varphi)\sqrt{-1}dz^i\wedge d\overline{z^i}\int_{t=0}^1(\omega + t\sqrt{-1}\partial\bar\partial\varphi)^{m-1}dt\\ &= m\, h'(\varphi)\sqrt{-1}dz^i\wedge d\overline{z^i}\int_{t=0}^1((1-t)\omega + t\omega')^{m-1}dt\\ &\geq m\, h'(\varphi)\sqrt{-1}dz^i\wedge d\overline{z^i}\int_{t=0}^1(1-t)^{m-1}\omega^{m-1}dt\\ &= h'(\varphi)\sqrt{-1}dz^i\wedge d\overline{z^i}\wedge\omega^{m-1}.\end{aligned}$$

Hence for $h'(\varphi) \geq 0$

$$\left|\int\left[(\omega + \sqrt{-1}\partial\bar{\partial}\varphi)^m - \omega^m\right]h(\varphi)\right| \geq \int h'(\varphi)\,|\partial\varphi|^2.$$

Take $\alpha \geq 0$. Let $h(x) = x|x|^\alpha$. Then $h'(x) = (\alpha+1)\,|x|^\alpha$. Thus

$$\left|\int\left[(\omega + \sqrt{-1}\partial\bar{\partial}\varphi)^m - \omega^m\right]\varphi|\varphi|^\alpha\right| \geq c(\alpha+1)\int|\varphi|^\alpha|\partial\varphi|^2 \tag{2.3.1}$$
$$= c\,\frac{\alpha + 1}{(\frac{\alpha}{2} + 1)^2}\int|\partial(\varphi|\varphi|^{\alpha/2})|^2.$$

This is the end of the argument corresponding to the process in the Moser iteration technique of multiplying the linear elliptic equation by a power of the solution and integrating by parts. As a result we now have an estimate of the L^2 norm of the derivative of a power of the solution in terms of its L^2 norm of a lower power of the solution, because the Monge-Ampére equation makes $(\omega + \sqrt{-1}\partial\bar{\partial}\varphi)^m$ a known entity. It is important to know that a *lower* power of the solution occurs in the estimate.

(2.4) We can now start the iteration. Since we have a compact domain, the argument is easier than Moser's orginal argument for a noncompact domain where the domain has to be shrunken *ever more slightly* each time in the infinite process to finally get the estimate for a smaller domain. We apply the Sobolev inequality [G-T, p.155, Th.7.10]

$$\|u\|_{L^r} \leq C'(\|\nabla u\|_{L^p} + \|u\|_{L^p})$$

with $r = \frac{2mp}{2m-p}$. In the case $p = 2$ we have

$$\|u\|_{\frac{2m}{m-1}} \leq C(\|\nabla u\|_2 + \|u\|_2).$$

We use the constant C in the generic sense. So C may mean different constants in different equations. Let $\beta = \frac{m}{m-1}$. From (2.3.1) we have

(2.4.1) $$\int |\partial(\varphi|\varphi|^{\alpha/2})|^2 \le C(\alpha+1)\int |\varphi|^{\alpha+1}.$$

Thus by using $u = \varphi|\varphi|^{\alpha/2}$ in the Sobolev inequality, we get

$$\begin{aligned}\left[\int |\varphi|^{(\alpha+2)\beta}\right]^{1/\beta} &\le C\left[\int |\partial(\varphi|\varphi|^{\alpha/2})|^2 + \int |\varphi|^{\alpha+2}\right] \\ &\le C\left[(\alpha+1)\int |\varphi|^{\alpha+1} + \int |\varphi|^{\alpha+2}\right] \\ &\le C\left[(\alpha+1)\left[\int |\varphi|^{\alpha+1}\right]^{\frac{\alpha+1}{\alpha+2}}(\mathrm{Vol}\ M)^{\frac{1}{\alpha+2}} + \int |\varphi|^{\alpha+2}\right].\end{aligned}$$

Let $p = \alpha + 2 \ge 2$. Then

$$\begin{aligned}\left[\int |\varphi|^{p\beta}\right]^{1/\beta} &\le Cp\left[1 + \int |\varphi|^p\right] \\ &\le 2Cp\ \mathrm{Max}(1, \int |\varphi|^p)\end{aligned}$$

and

$$\mathrm{Max}(1, \|\varphi\|_{p\beta}) \le C^{1/p}p^{1/p}\ \mathrm{Max}(1, \|\varphi\|_p).$$

Hence

$$\log \mathrm{Max}(1, \|\varphi\|_{p\beta}) \le \frac{1}{p}\log C + \frac{1}{p}\log p + \log \mathrm{Max}(1, \|\varphi\|_p).$$

Successively replacing p by $p\beta$ and summing up, we have

$$\begin{aligned}\log \mathrm{Max}(1, \|\varphi\|_{p\beta^k}) &\le \frac{1}{p}\left(\Sigma_{i=0}^{k-1}\frac{1}{\beta^i}\right)\log C + \frac{1}{p}\left(\Sigma_{i=0}^{k-1}\frac{1}{\beta^i}\right)\log p \\ &\quad + \frac{1}{p}\left(\Sigma_{i=0}^{k-1}\frac{i}{\beta^i}\right)\log \beta + \log \mathrm{Max}(1, \|\varphi\|_p).\end{aligned}$$

Let $p = 2$ and $k \to \infty$, we get

$$\log \operatorname{Max}(1, \|\varphi\|_\infty) \leq C + \log \operatorname{Max}(1, \|\varphi\|_2).$$

(2.5) To finish our zeroth order estimate, it suffices to estimate $\|\varphi\|_2$. When $\alpha = 0$ the inequality (2.4.1) yields

$$\int |\nabla\varphi|^2 \leq C \int |\varphi|.$$

Since $\int \varphi = 0$, φ is orthogonal to the kernel of Δ. Let λ_1 be the first nonzero eigenvalue of $-\Delta$. Then $-(\Delta\psi,\psi) \geq \lambda_1(\psi,\psi)$ for ψ perpendicular to the kernel of Δ. That is, $(\nabla\psi,\nabla\psi) \geq \lambda_1(\psi,\psi)$ for ψ perpendicular to the kernel of Δ. So

$$\int |\varphi|^2 \leq \frac{1}{\lambda_1} \int |\nabla\varphi|^2$$

and

$$\int |\varphi|^2 \leq C \int |\varphi| \leq C \left[\int |\varphi|^2\right]^{1/2} (\mathrm{Vol}\ M)^{1/2}$$

and $\int |\varphi|^2 \leq C \,\mathrm{Vol}\ M$.

(2.6) The above argument of integration by parts and iteration to get the zeroth order *a priori* estimates actually yields the following. Suppose θ is a Kähler form on a compact complex manifold of complex dimension m and ψ is a smooth real-valued function on the manifold. If the supremum norm of $(\theta + \sqrt{-1}\partial\bar{\partial}\psi)^m$ is bounded by some constant C, then the oscillation of ψ (*i.e.* $\sup \psi - \inf \psi$) on the manifold is bounded by some constant depending only on C and the manifold.

§3. *Second Order Estimates.*

(3.1) We now do the second order *a priori* estimate of φ. Since $g_{i\bar{j}} + \partial_i \partial_{\bar{j}} \varphi$ is positive definite, to get a second order estimate of φ it suffices to have an upper bound estimate of $m + \Delta\varphi$. We want to do this by the maximum principle. So we want to consider some elliptic inequality satisfied by $m + \Delta\varphi$. There are two natural elliptic operators. One is Δ and the other is Δ'. We do not know yet which one to use. In any case the inequality involves the fourth order derivatives of φ. Let us differentiate φ four times before we decide. The only equation we have involves the Ricci curvature R'_{ij} of the new metric. So we should express R'_{ij} in terms of the fourth order derivatives of φ. To make the computation easier we use normal coordinates. So we fix a point and choose normal coordinates at that point for $g_{i\bar{j}}$ so that $g'_{i\bar{j}}$ is also diagonal. Applying $-\partial_k \partial_{\bar{\ell}}$ to the equation $g'_{i\bar{j}} = g_{i\bar{j}} + \varphi_{i\bar{j}}$ (where $\varphi_{i\bar{j}}$ means $\partial_i \partial_{\bar{j}} \varphi$), we get

$$- \partial_k \partial_{\bar{\ell}} g'_{i\bar{j}} = R_{i\bar{j}k\bar{\ell}} - \partial_k \partial_{\bar{\ell}} \varphi_{i\bar{j}}$$

and

$$R'_{i\bar{j}k\bar{\ell}} = g'^{p\bar{q}} \partial_i g'_{k\bar{q}} \partial_{\bar{j}} g'_{p\bar{\ell}} + R_{i\bar{j}k\bar{\ell}} - \partial_k \partial_{\bar{\ell}} \varphi_{i\bar{j}}.$$

Now to get the new Ricci tensor we should contract this equation with $g'^{k\bar{\ell}}$. Since we want to get $\Delta\varphi$, we should also contract it with $g^{i\bar{j}}$. These two contractions yield

$$g^{i\bar{j}} R'_{ij} = g^{i\bar{j}} g'^{k\bar{\ell}} g'^{p\bar{q}} \partial_i g'_{k\bar{q}} \partial_{\bar{j}} g'_{p\bar{\ell}} + g'^{k\bar{\ell}} R_{k\bar{\ell}} - g^{i\bar{j}} g'^{k\bar{\ell}} \partial_k \partial_{\bar{\ell}} \varphi_{i\bar{j}}.$$

Note that this equation is only a consequence of the equation $g'_{i\bar{j}} = g_{i\bar{j}} + \varphi_{i\bar{j}}$ and has nothing to do yet with the Monge-Ampère equation. This equation suggests that for the elliptic inequality we should use $\Delta'(m + \Delta\varphi)$. We should express the term involving the fourth order derivative of φ in terms of $\Delta'(m + \Delta\varphi)$.

$$\begin{aligned} g^{i\bar{j}} g'^{k\bar{\ell}} \partial_k \partial_{\bar{\ell}} \varphi_{i\bar{j}} &= g'^{k\bar{\ell}} \partial_k \partial_{\bar{\ell}} (g^{i\bar{j}} \varphi_{i\bar{j}}) - g'^{k\bar{\ell}} R^{i\bar{j}}{}_{k\bar{\ell}} \varphi_{i\bar{j}} \\ &= g'^{k\bar{\ell}} \partial_k \partial_{\bar{\ell}} (g^{i\bar{j}} \varphi_{i\bar{j}}) - g'^{k\bar{\ell}} R^{i\bar{j}}{}_{k\bar{\ell}} g'_{i\bar{j}} + g'^{k\bar{\ell}} R_{k\bar{\ell}} \\ &= \Delta'(m + \Delta\varphi) - g'^{k\bar{\ell}} R^{i\bar{j}}{}_{k\bar{\ell}} g'_{i\bar{j}} + g'^{k\bar{\ell}} R_{k\bar{\ell}}. \end{aligned}$$

Thus

$$g^{i\bar{j}} R'_{i\bar{j}} = g^{i\bar{j}} g'^{k\bar{\ell}} g'^{p\bar{q}} \partial_i g'_{k\bar{q}} \partial_{\bar{j}} g'_{p\bar{\ell}} - \Delta'(m + \Delta\varphi) + g'^{k\bar{\ell}} R^{i\bar{j}}{}_{k\bar{\ell}} g'_{i\bar{j}}.$$

Now we use $R'_{i\bar{j}} = R_{i\bar{j}} + c\varphi_{i\bar{j}} - tF_{i\bar{j}}$ and get

$$\Delta'(m + \Delta\varphi) = -g^{i\bar{j}} R_{i\bar{j}} - c\Delta\varphi + t\Delta F + g^{i\bar{j}} g'^{k\bar{\ell}} g'^{p\bar{q}} \partial_i g'_{k\bar{q}} \partial_{\bar{j}} g'_{p\bar{\ell}} + g'^{k\bar{\ell}} R^{i\bar{j}}{}_{k\bar{\ell}} g'_{i\bar{j}}.$$

(3.2) Everything is in order for the application of the maximum principle to get an upper bound estimate of $m + \Delta\varphi$ except the term $g'^{k\bar{\ell}} R^{i\bar{j}}{}_{k\bar{\ell}} g'_{i\bar{j}}$. This term can be rewritten as

$$g'^{k\bar{\ell}} R^{i\bar{j}}{}_{k\bar{\ell}} g'_{i\bar{j}} = \Sigma_{i,j} \frac{1 + \varphi_{i\bar{i}}}{1 + \varphi_{j\bar{j}}} R_{i\bar{i}j\bar{j}}.$$

Its absolute value is dominated by $(m+\Delta\varphi)\, \Sigma_j \frac{1}{1 + \varphi_{j\bar{j}}}$. To take care of this term, we consider $\Delta'\log(m+\Delta\varphi)$ instead of $\Delta'(m+\Delta\varphi)$ to get rid of the factor $(m+\Delta\varphi)$ in $(m+\Delta\varphi)\, \Sigma_j \frac{1}{1 + \varphi_{j\bar{j}}}$. To get an inequality for $\Delta'\log(m+\Delta\varphi)$ first we observe that by Hölder inequality we have

$$g^{i\bar{j}} g'^{k\bar{\ell}} g'^{p\bar{q}} \partial_i g'_{k\bar{q}} \partial_{\bar{j}} g'_{p\bar{\ell}} \geq \frac{|\nabla'(m+\Delta\varphi)|^2}{m+\Delta\varphi},$$

where $|\nabla'\tau|^2$ means $g'^{i\bar{j}}\partial_i\tau\partial_{\bar{j}}\bar{\tau}$, because

$$\begin{aligned}
|\nabla'(m+\Delta\varphi)|^2 &= \Sigma_{i,j,k}\frac{1}{1+\varphi_{k\bar{k}}}\partial_k\varphi_{i\bar{i}}\partial_{\bar{k}}\varphi_{j\bar{j}} \\
&= \Sigma_{i,j}\left[\Sigma_k\frac{1}{\sqrt{1+\varphi_{k\bar{k}}}}\partial_k\varphi_{i\bar{i}}\frac{1}{\sqrt{1+\varphi_{k\bar{k}}}}\partial_k\varphi_{j\bar{j}}\right] \\
&\le \Sigma_{i,j}\left[\Sigma_k\frac{1}{1+\varphi_{k\bar{k}}}|\partial_k\varphi_{i\bar{i}}|^2\right]^{\frac{1}{2}}\left[\Sigma_k\frac{1}{1+\varphi_{k\bar{k}}}|\partial_k\varphi_{j\bar{j}}|^2\right]^{\frac{1}{2}} \\
&= \left[\Sigma_i\left[\Sigma_k\frac{1}{1+\varphi_{k\bar{k}}}|\partial_k\varphi_{i\bar{i}}|^2\right]^{\frac{1}{2}}\right]^2 \\
&= \left[\Sigma_i\sqrt{1+\varphi_{i\bar{i}}}\left[\Sigma_k\frac{1}{1+\varphi_{i\bar{i}}}\frac{1}{1+\varphi_{k\bar{k}}}|\partial_k\varphi_{i\bar{i}}|^2\right]^{\frac{1}{2}}\right]^2 \\
&\le \left[\Sigma_i(1+\varphi_{i\bar{i}})\right]\left[\Sigma_{i,k}\frac{1}{1+\varphi_{i\bar{i}}}\frac{1}{1+\varphi_{k\bar{k}}}|\partial_k\varphi_{i\bar{i}}|^2\right] \\
&\le (m+\Delta\varphi)\,\Sigma_{i,k,p}\frac{1}{1+\varphi_{k\bar{k}}}\frac{1}{1+\varphi_{p\bar{p}}}\partial_i\varphi_{k\bar{p}}\partial_{\bar{i}}\varphi_{p\bar{k}}
\end{aligned}$$

(after we use $\partial_i\varphi_{k\bar{p}} = \partial_k\varphi_{i\bar{p}}$). Thus we have

$$\begin{aligned}
\Delta'\log(m+\Delta\varphi) &= \frac{\Delta'(m+\Delta\varphi)}{m+\Delta\varphi} - \frac{|\nabla'(m+\Delta\varphi)|^2}{(m+\Delta\varphi)^2} \\
&\ge \frac{1}{m+\Delta\varphi}\left[-g^{i\bar{j}}R_{i\bar{j}} - c\Delta\varphi + t\Delta F + g'^{k\bar{\ell}}R^{i\bar{j}}{}_{k\bar{\ell}}g'_{i\bar{j}}\right] \\
&\ge -C\left(1+\frac{1}{m+\Delta\varphi}+\Sigma_j\frac{1}{1+\varphi_{j\bar{j}}}\right) \\
&\ge -C'\left(1+\Sigma_j\frac{1}{1+\varphi_{j\bar{j}}}\right).
\end{aligned}$$

(3.3) To use the maximum principle we have to take care of the term $\Sigma_j\frac{1}{1+\varphi_{j\bar{j}}}$. The idea is look for some estimable function whose Δ' dominates $C'\Sigma_j\frac{1}{1+\varphi_{j\bar{j}}}$ and add that function to $m+\Delta\varphi$ before we apply the maximum principle. A good candidate is a multiple of the function φ, because

$$\Delta'\varphi = \Sigma_j\frac{\varphi_{j\bar{j}}}{1+\varphi_{j\bar{j}}} = m - \Sigma_j\frac{1}{1+\varphi_{j\bar{j}}}.$$

By applying the maximum principle to the elliptic inequality

$$\Delta'((m+\Delta\varphi) + (C'+1)\varphi) \geq -C' + \Sigma_j \frac{1}{1+\varphi_{j\bar{j}}}.$$

At the point where the maximum of $(m+\Delta\varphi) + (C'+1)\varphi$ is achieved, we have $\Sigma_j \frac{1}{1+\varphi_{j\bar{j}}} \leq C'$, which implies that

$$(m+\Delta\varphi)A_t e^{c\varphi-tF} = (m+\Delta\varphi)\left[\Pi_j \frac{1}{1+\varphi_{j\bar{j}}}\right] \leq \left[\Sigma_j \frac{1}{1+\varphi_{j\bar{j}}}\right]^{m-1} \leq C'^{m-1},$$

because trivially $\Sigma_{\nu=1}^{m} a_1 \cdots \hat{a}_\nu \cdots a_m \leq (\Sigma_{\nu=1}^{m} a_\nu)^{m-1}$ for any positive a_ν $(1 \leq \nu \leq m)$, where $\hat{a}_\nu$ means that a_ν is omitted. Thus the maximum of $m+\Delta\varphi$ is dominated by $\text{constant}\cdot\exp(\sup(-c\varphi + tF))$. Since

$$\Sigma_j \frac{1}{1+\varphi_{j\bar{j}}} \leq (m+\Delta\varphi)^{m-1}\left[\Pi_j \frac{1}{1+\varphi_{j\bar{j}}}\right] = (m+\Delta\varphi)^{m-1} A_t e^{c\varphi-tF},$$

we have the upper bound

$$\text{constant}\cdot\exp((m-1)\sup(-c\varphi + tF))\exp(\sup(c\varphi - tF))$$

for $\Sigma_j \frac{1}{1+\varphi_{j\bar{j}}}$.

§4. *Hölder estimates for the second derivatives.*

(4.1) Let u be a real-valued function on an open subset Ω of $\mathbb{C}^m$. Consider the Monge-Ampème equation $\log\det(\partial_i\partial_{\bar{j}}u) = h$. Here $u = \psi + \varphi$ when $\partial_i\partial_{\bar{j}}\psi = g_{i\bar{j}}$ and $g = cu + F - \psi + \log\det(g_{i\bar{j}})$. Let $\Phi(D\bar{D}u) = \log\det(\partial_i\partial_{\bar{j}}u)$.

Let γ be an arbitrary vector of $\mathbb{R}^n = \mathbb{C}^m$. Differentiating $\phi(D\bar{D}u) = h$ with respect to γ and then with respect to $\bar{\gamma}$ gives

$$\frac{\partial\phi}{\partial u_{i\bar{j}}} u_{i\bar{j}\gamma} = h_\gamma$$

(4.1.1) $$\frac{\partial^2\phi}{\partial u_{i\bar{j}}\partial u_{k\bar{\ell}}} u_{k\bar{\ell}\bar{\gamma}} u_{i\bar{j}\gamma} + \frac{\partial\phi}{\partial u_{i\bar{j}}} u_{i\bar{j}\gamma\bar{\gamma}} = h_{\gamma\bar{\gamma}}.$$

Here the summation convention of summing over repeated indices is being used. Now we observe that ϕ is concave as a function of $(u_{i\bar{j}})$. To see this concavity, we take two matrices $(A_{i\bar{j}})$ and $(B_{i\bar{j}})$ and diagonalize *as Hermitian forms (and not as Hermitian matrices)* and get eigenvalues λ_i and μ_i $(1 \leq i \leq m)$. Concavity means that

$$t\phi(B) + (1-t)\phi(A) \leq \phi(tB + (1-t)A)$$

for $0 \leq t \leq 1$. We have $\phi(A) = \log\det A = \log(\lambda_1 \cdots \lambda_m) = \Sigma_{i=1}^m \log\lambda_i$ and $\phi(B) = \log\det B = \log(\mu_1 \cdots \mu_m) = \Sigma_{i=1}^m \log\mu_i$. Hence

$$\begin{aligned} t\phi(B) + (1-t)\phi(A) \leq \phi(tB + (1-t)A) &= t\Sigma_{i=1}^m \log\mu_i + (1-t)\Sigma_{i=1}^m \log\lambda_i \\ &= \Sigma_{i=1}^m (t\log\mu_i + (1-t)\log\lambda_i) \\ &\leq \Sigma_{i=1}^m \log(t\,\mu_i + (1-t)\,\lambda_i) \quad \text{(by concavity of log)} \\ &= \phi(tB + (1-t)A) \end{aligned}$$

So the matrix $\left[\frac{\partial^2\phi}{\partial u_{i\bar{j}}\partial u_{k\bar{\ell}}}\right]$ is seminegative as a matrix in $(u_{i\bar{j}})$.

Thus from (4.1.1) we conclude that $\frac{\partial\phi}{\partial u_{i\bar{j}}} u_{i\bar{j}\gamma\bar{\gamma}} \geq h_{\gamma\bar{\gamma}}$. Let $w = D_{\gamma\bar{\gamma}}u$. We rewrite the equation as $g^{i\bar{j}}\partial_i\partial_{\bar{j}}w \geq h_{\gamma\bar{\gamma}}$.

(4.2) We need the following Harnack inequality for a linear elliptic equation $g^{i\bar{j}}\partial_i\partial_{\bar{j}}v \leq \theta$ with $v \geq 0$ on the ball B_{2R} of radius $2R$ in $\mathbb{C}^m$ centered at 0. Take $q > m$. By the Harnack inequality whose derivation we will do later, there exist constants $p > 0$ and $C > 0$ such that

$$\left[\frac{1}{R^{2m}}\int_{B_R} v^p\right]^{1/p} \leq C\left[\inf_{B_R} v + R^{\frac{2(q-m)}{q}} \|\theta\|_{L^q(B_{2R})}\right].$$

Let us assume this Harnack inequality and finish with our Hölder estimates of the second derivative. For $s = 1,2$ let $M_s = \sup_{B_{sR}} w$. Applying the Harnack inequality to $M_2 - w$, we get

$$\text{(4.2.1)} \quad \left[\frac{1}{R^{2m}}\int_{B_R} (M_2 - w)^p\right]^{1/p} \leq C\left[M_2 - M_1 + R^{\frac{2(q-m)}{q}} \|D_{\gamma\bar{\gamma}}h\|_{L^q(B_{2R})}\right].$$

At this point an ingenious trick has to be used. In order to be able to reduce the order of the elliptic differential equation (4.1.1) by the substitution $w = D_{\gamma\bar{\gamma}}u$, we use the concavity of the function ϕ and end up with an elliptic inequality instead of an elliptic equation. As a consequence we have the estimate (4.2.1) for the supersolution of an elliptic equation. However, because we have an elliptic inequality instead of an elliptic equation, we do not have the estimate corresponding to the *subsolution* of an elliptic equation. We need both estimates to get the required Hölder estimate. We are going to use the concavity of the function ϕ to compensate for this.

By the concavity of ϕ in $u_{i\bar{j}}$ the tangent plane to the graph of ϕ at the point $(D_{i\bar{j}}u(y))$ is above the graph of ϕ. Since the equation of the tangent plane to the graph of F at the point $(D_{i\bar{j}}u(y))$ is $\phi(D\bar{D}u(y)) + \phi_{i\bar{j}}(D\bar{D}u)(D_{i\bar{j}}u(x) - D_{i\bar{j}}u(y))$ in the variable $(D_{i\bar{j}}u(x))$, it follows that

$$\phi(D\bar{D}u(y)) + \phi_{i\bar{j}}(D\bar{D}u(y))(D_{i\bar{j}}u(x) - D_{i\bar{j}}u(y)) \geq \phi(D\bar{D}u(x)).$$

In other words,

$$\begin{aligned}\phi_{i\bar{j}}(D\bar{D}u(y))(D_{i\bar{j}}u(x) - D_{i\bar{j}}u(y)) &\geq \phi(D\bar{D}u(x)) - \phi(D\bar{D}u(y)) \\ &= h(x) - h(y).\end{aligned}$$

Changing the signs of both sides we have

$$\phi_{i\bar{j}}(D\bar{D}u(y))(D_{i\bar{j}}u(y) - D_{i\bar{j}}u(x)) \leq h(y) - h(x).$$

(4.3) We now need a lemma in simple linear algebra. Let $S(\lambda,\Lambda)$ be the set of all $m\times m$ positive matrices with complex number entries whose eigenvalues are between λ and Λ. Then there exist a finite number of unit vectors $\gamma_1,\cdots,\gamma_N$ in $\mathbb{C}^m$ and λ^* and Λ^* depending only on m, λ and Λ such that any matrix $A = (a_{i\bar{j}})$ can be written as $a_{i\bar{j}} = \Sigma_{k=1}^{N}\beta_k\gamma_{ki}\overline{\gamma_{kj}}$ with $\lambda^*\leq \beta_k \leq \Lambda^*$ $(k = 1,\cdots,N)$ where $\gamma_k = (\gamma_{k1},\cdots,\gamma_{km})$. The proof of the lemma is as follows. Every matrix in $S(\frac{\lambda}{2},\Lambda)$ can be written as $\Sigma_{\upsilon=1}^{m}\beta_\upsilon\gamma_\upsilon\otimes\overline{\gamma_\upsilon}$ with $\frac{\lambda}{2}\leq\beta\leq\Lambda$, where γ_υ is a unit vector. By compactness we can cover $S(\frac{\lambda}{2},\Lambda)$ by a finite number of open sets of the form

$$U(\gamma_1,\cdots,\gamma_{m^2(m+1)}) = \{\Sigma_{\upsilon=1}^{m^2(m+1)}\beta_\upsilon\gamma_\upsilon\otimes\overline{\gamma_\upsilon} \,\big|\, 2\Lambda > \beta_\upsilon > 0\}.$$

Here we use $m^2(m+1)$ vectors $\gamma_1,\cdots,\gamma_{m^2(m+1)}$ to make sure that we do have an open subset of the set of all $m\times m$ positive matrices with complex number entries, because the dimension of the ambient space is $\frac{1}{2}m(m+1)$ over $\mathbb{C}$ and we need $m(m+1)$ matrices to span an open *convex* set over $\mathbb{R}$. Each matrix takes up m unit vectors γ_k and we have $m(m+1)$ such matrices. The bound $2\Lambda > \beta_\upsilon$ is assured by the fact that we are considering a convex subset. So we need $m^2(m+1)$ such γ_k. Thus we can find $\gamma_1, \cdots, \gamma_N$ so that each matrix A in $S(\frac{\lambda}{2}, \Lambda)$ can be written as $\Sigma_{\upsilon=1}^{N} \beta_\upsilon \gamma_\upsilon \otimes \overline{\gamma_\upsilon}$ with $\beta_\upsilon > 0$. Take a matrix A in $S(\lambda, \Lambda)$. then $A - \Sigma_{\upsilon=1}^{N} \frac{\lambda}{2N} \gamma_\upsilon \otimes \overline{\gamma_\upsilon}$ belongs to $S(\frac{\lambda}{2}, \Lambda)$. So we have $A = \Sigma_{\upsilon=1}^{N} (\beta_\upsilon + \frac{\lambda}{2N}) \gamma_\upsilon \otimes \overline{\gamma_\upsilon}$ with $\beta_\upsilon + \frac{\lambda}{2N} \geq \frac{\lambda}{2N} = \lambda^*$. We can also assume that the set $\gamma_1, \cdots, \gamma_N$ contains an orthonormal basis by throwing in such a basis in $\gamma_1, \cdots, \gamma_N$.

(4.4) We now apply the linear algebra lemma to the matrix $(\Phi_{i\bar{j}}(D\bar{D}u(y)))$. Because we already have the *a priori* second order estimates of φ, the matrix $(\Phi_{i\bar{j}}(D\bar{D}u(y)))$ belongs to $S(\lambda, \Lambda)$. So the matrix $(\Phi_{i\bar{j}}(D\bar{D}u(y)))$ is of the form $\Sigma_{\upsilon=1}^{m} \beta_\upsilon \gamma_\upsilon \otimes \overline{\gamma_\upsilon}$ with $\lambda^* \leq \beta_k \leq \Lambda^*$ $(k = 1, \cdots, N)$. Let $w_\upsilon = D_{\gamma_\upsilon \overline{\gamma_\upsilon}} u$. Then $w_\upsilon = \Sigma_{i,j=1}^{m} \gamma_{\upsilon i} \overline{\gamma_{\upsilon j}} D_{i\bar{j}} u$. Thus

$$\begin{aligned}\Phi_{i\bar{j}}(D\bar{D}u(y))(D_{i\bar{j}}u(y) - D_{i\bar{j}}u(x)) &= \Sigma_{\upsilon=1}^{m} \beta_\upsilon \gamma_{\upsilon i} \overline{\gamma_{\upsilon j}} (D_{i\bar{j}}u(y) - D_{i\bar{j}}u(x)) \\ &= \Sigma_{\upsilon=1}^{m} \beta_\upsilon (w_\upsilon(y) - w_\upsilon(x))\end{aligned}$$

and

$$\Sigma_{\upsilon=1}^{m} \beta_\upsilon (w_\upsilon(y) - w_\upsilon(x)) \leq h(y) - h(x). \tag{4.4.1}$$

We apply the Harnack inequality to the case $\gamma = \gamma_\upsilon$ $(1 \leq \upsilon \leq N)$. For $1 \leq \upsilon \leq N$ and $s = 1, 2$ let $M_{s\upsilon} = \sup_{B_{sR}} w_\upsilon$ and $m_{s\upsilon} = \inf_{B_{sR}} w_\upsilon$. We get

$$\begin{aligned}&\left[\frac{1}{R^n} \int_{B_R} (\Sigma_{k\neq\ell}(M_{2k} - w_k))^p\right]^{1/p} \\ &\leq N^{1/p} \Sigma_{k\neq\ell} \left[\frac{1}{R^n} \int_{B_R} (M_{2k} - w_k)^p\right]^{1/p} \\ &\leq C\left[\Sigma_{k\neq\ell}(M_{2k} - M_{1k}) + R^2 \sup_{B_{2R}} |D^2 g|\right]\end{aligned} \tag{4.4.2}$$

$$\leq C\left[\omega(2R) - \omega(R) + R^2 \sup_{B_{2R}} |D^2 g|\right].$$

where

$$\omega(sR) = \Sigma_{k=1}^{N} \operatorname{osc}_{B_{sR}} w_k = \Sigma_{k=1}^{N} (M_{sk} - m_{sk}),$$

because

$$(M_{2k} - m_{2k}) - (M_{1k} - m_{1k}) = (M_{2k} - M_{1k}) + (m_{1k} - m_{2k}) \geq M_{2k} - M_{1k}.$$

From (4.4.1) we have

$$\beta_\ell (w_\ell(y) - w_\ell(x)) \leq h(y) - h(x) + \Sigma_{k\neq\ell}\, \beta_k (w_k(x) - w_k(y)).$$

Hence by choosing x so that $w_\ell(x)$ approaches $m_{2\ell}$ and using the mean value theorem we have

$$w_\ell(y) - m_{2\ell} \leq \frac{1}{\lambda^*} \{3R \sup_{B_{2R}} |Dh| + \Lambda^* \Sigma_{k\neq\ell} (M_{2k} - w_k(y))\}.$$

After integrating over $y \in B_R$ and using (4.4.2) we have

(4.4.3)
$$\left[\frac{1}{R^{2m}} \int_{B_R} (w_\ell(y) - m_{2\ell})^p\right]^{1/p} \leq C\left[\omega(2R) - \omega(R) + R \sup_{B_{2R}} |Dh| + R^2 \sup_{B_{2R}} |D^2 h|\right].$$

This is the estimate corresponding to the subsolution of an elliptic equation. The main idea of the preceding argument is that in (4.4.1) we have an estimate of the sum in the direction we want. This couples with the estimate of each summand in the opposite direction gives us the required estimate of each summand in the direction we want.

(4.5) We now combine the estimates corresponding to both the supersolution and the subsolution of an elliptic equation and use the method of Moser [Mo1] to get Hölder estimates. From (4.2.1) we have

$$\left[\frac{1}{R^n}\int_{B_R}(M_{2\ell}-w_\ell)^p\right]^{1/p} \le C\left[M_{2\ell}-M_{1\ell}+R^2\sup_{B_{2R}}|D^2h|\right].$$

Adding it to (4.4.3), we get

$$M_{2\ell}-m_{2\ell} \le C\left[\omega(2R)-\omega(R)+R\,\sup_{B_{2R}}|Dh|+R^2\sup_{B_{2R}}|D^2h|\right].$$

Summing over ℓ from 1 to N we have

$$\omega(2R) \le C\left[\omega(2R)-\omega(R)+R\,\sup_{B_{2R}}|Dh|+R^2\sup_{B_{2R}}|D^2h|\right].$$

So

$$\omega(R) \le \delta\omega(2R)+R\,\sup_{B_{2R}}|Dh|+R^2\sup_{B_{2R}}|D^2h|,$$

where $\delta = 1-\frac{1}{C}$. This is true for all R. Hence we have the Hölder estimate of D^2u from the following lemma.

(4.6) Suppose τ and γ are positive numbers less than 1. Let ω be a nondecreasing function on $(0,R_0]$ and $\omega(\tau R) \le \gamma\omega(R)+\sigma(R)$. Then for any $0 < \nu < 1$ and $R \le R_0$ one has $\omega(R) \le C\left[\left(\frac{R}{R_0}\right)^\alpha \omega(R_0)+\sigma(R^\mu R_0^{1-\mu})\right]$, where $C = C(\gamma,\tau)$ and $\alpha = \alpha(\gamma,\tau,\mu)$ are positive constants. To see this, fix $R_1 < R_0$. Then $\omega(\tau^p R_1) \le \gamma^p\omega(R_1)+\sigma(R_1)\Sigma_{i=0}^{p-1}\gamma^i \le \gamma^p\omega(R_1)+\frac{\sigma(R_1)}{1-\gamma}$. For any $R \le R_1$ there exists a unique p such that $\tau^pR_1 < R < \tau^{p-1}R_1$. So $\tau^p < \frac{R}{R_1}$

and $p < \frac{\log(\frac{R}{R_1})}{\log \tau}$. Hence

$$\omega(R) \le \omega(\tau^p R_1) \le \gamma^p \omega(R_0) + \frac{\sigma(R_1)}{1-\gamma}$$
$$\le \frac{1}{\gamma} \gamma^{\frac{\log(\frac{R}{R_1})}{\log \gamma}} \omega(R_0) + \frac{\sigma(R_1)}{1-\gamma}$$
$$\le \frac{1}{\gamma} \exp\left[\log \gamma \frac{\log(\frac{R}{R_1})}{\log \tau}\right] \omega(R_0) + \frac{\sigma(R_1)}{1-\gamma}$$
$$\le \frac{1}{\gamma} (\frac{R}{R_1})^{\frac{\log \gamma}{\log \tau}} + \frac{\sigma(R_1)}{1-\gamma}.$$

Finally we choose $R_1 = R^\mu R_0^{1-\mu}$ and get

$$\omega(R) \le \left[\frac{1}{\gamma} R_0^{-(1-\mu)\frac{\log \gamma}{\log \tau}}\right] R^{(1-\mu)\frac{\log \gamma}{\log \tau}} + \frac{1}{1-\gamma} \sigma(R^\mu R_0^{1-\mu}).$$

Now we turn to the Harnack inequality.

§5 *Derivation of Harnack inequality by Moser's iteration technique.*

(5.1) We now prove the following Harnack inequality. Let B_R denote the open ball of radius R in $\mathbb{C}^m$ centered at the origin. Suppose that on B_{2R} we have the inequality $g^{\alpha\bar{\beta}} \partial_\alpha \partial_{\bar{\beta}} v \le \theta$ for some smooth real-valued functions v and θ, where $g_{\alpha\bar{\beta}}$ is a Kähler metric on B_{2R} so that both the matrix $(g_{\alpha\bar{\beta}})$ and its inverse $(g^{\alpha\bar{\beta}})$ have *a priori* positive bounds. Assume that v is positive on B_{2R}. Let $q > m$. The Harnack inequality we want to prove is the following. There exist positive number p such that

$$R^{-2n/m}\|v\|_{L^p(B_{2R})} \leq C\left[\inf_{B_R} v + R^{\frac{2(q-m)}{q}} \|\theta\|_{L^q(B_{2R})}\right],$$

where C is a constant depending only on m, p, q and the *a priori* positive bounds of $(g_{\alpha\bar{\beta}})$ and $(g^{\alpha\bar{\beta}})$.

First assume by rescaling that $R = 1$. Let A be the L^q norm of θ over the ball B_2. Let $w = v+A$. Take $\upsilon > 0$. Then $\partial_{\bar{\beta}} w^{-\upsilon} = -\upsilon(v+A)^{-\upsilon-1}\partial_{\bar{\beta}}u$ and

$$\begin{aligned} g^{\alpha\bar{\beta}}\partial_\alpha\partial_{\bar{\beta}} w^{-\upsilon} &= -\upsilon(v+A)^{-\upsilon-1}g^{\alpha\bar{\beta}}\partial_\alpha\partial_{\bar{\beta}}u + \upsilon(\upsilon+1)(v+A)^{-\upsilon-2}g^{\alpha\bar{\beta}}(\partial_\alpha v)(\partial_{\bar{\beta}}v) \\ &\geq -\upsilon\,\theta\,(v+A)^{-\upsilon-1} \geq -\upsilon\,\xi\,w^{-\upsilon}, \end{aligned}$$

where $\xi = \frac{\theta}{v+A}$. The L^q norm of ξ over B_2 is no more than $(\mathrm{Vol}\ B_2)^{1/q}$. Let us now introduce a cut-off function η. Multiply both sides by $\eta^2 w^{-\upsilon}$ and integrate over B_2 *with respect to the volume form of the Kähler metric* $g_{i\bar{j}}$ and we get

$$\int \eta^2 w^{-\upsilon} g^{\alpha\bar{\beta}}\partial_\alpha\partial_{\bar{\beta}} w^{-\upsilon} \geq -\int \upsilon\,\xi\eta^2 w^{-2\upsilon}$$

and

$$\int \eta^2 g^{\alpha\bar{\beta}}\partial_\alpha w^{-\upsilon}\partial_{\bar{\beta}}w^{-\upsilon} + \int 2\eta w^{-\upsilon}g^{\alpha\bar{\beta}}\partial_\alpha\eta\partial_{\bar{\beta}}w^{-\upsilon} \leq \int \upsilon\xi\eta^2 w^{-2\upsilon}.$$

Thus

$$\|\eta D w^{-\upsilon}\|^2_{L^2} \leq C(\|w^{-\upsilon}D\eta\|^2_{L^2} + \|\upsilon\xi\eta^2 w^{-2\upsilon}\|_{L^1}).$$

Or $\|D(\eta w^{-\upsilon})\|^2_{L^2} \leq C(\|w^{-\upsilon}D\eta\|^2_{L^2} + \|\upsilon h\eta^2 w^{-2\upsilon}\|_{L^1})$. Since $\eta w^{-\upsilon}$ has compact support in B_2, by Sobolev lemma we have

$$\|\eta w^{-\upsilon}\|^2_{L^{2m/(m-1)}} \le C(\|w^{-\upsilon}D\eta\|^2_{L^2} + \|\upsilon h\eta^2 w^{-2\upsilon}\|_{L^1}).$$

(5.2) Now

$$\|h\eta^2 w^{-2\upsilon}\|_{L^1} \le \|h\|_{L^q}\|\eta w^{-\upsilon}\|^2_{L^{2q/(q-1)}}.$$

By Hölder' inequality

$$\|\eta w^{-\upsilon}\|_{L^{2q/(q-1)}} \le \left[\|\eta w^{-\upsilon}\|_{L^{2m/(m-1)}}\right]^{\frac{m}{q}}\left[\|\eta w^{-\upsilon}\|_{L^2}\right]^{1-\frac{m}{q}}$$

because $\frac{q-1}{2q} = (1-\frac{m}{q})\frac{1}{2} + \frac{m}{q}\left[\frac{m-1}{2m}\right]$. Since

$$a^{\frac{m}{q}}\, b^{1-\frac{m}{q}} = (\epsilon\ a)^{\frac{m}{q}}(\epsilon^{\frac{m}{m-q}}b)^{1-\frac{m}{q}} \le \frac{m}{q}\ \epsilon\ a + (1-\frac{m}{q})\epsilon^{\frac{m}{m-q}}\ b \le \epsilon\ a + \epsilon^{\frac{m}{m-q}}\ b$$

for any positive numbers a, b and ϵ, it follows that

$$\|\eta w^{-\upsilon}\|^2_{L^{2q/(q-1)}} \le 2\epsilon^2\ \|\eta w^{-\upsilon}\|^2_{L^{2m/(m-1)}} + 2\epsilon^{\frac{2m}{m-q}}\ \|\eta w^{-\upsilon}\|^2_{L^2}$$

for any positive ϵ. Choose ϵ so that $C(\mathrm{Vol}\ B_2)^{1/q}\upsilon 2\epsilon^2 = \frac{1}{2}$. Since $\|\xi\|_{L^q(B_2)} \le (\mathrm{Vol}\ B_2)^{1/q}$, we have

$$\|\eta w^{-\upsilon}\|^2_{L^{2m/(m-1)}} \le 2C\ \|w^{-\upsilon}D\eta\|^2_{L^2} + (4C(\mathrm{Vol}\ B_2)^{1/q}\upsilon)^{\frac{q}{2(q-m)}}\ \|\eta w^{-\upsilon}\|^2_{L^2}.$$

Thus

$$\|w^{-\upsilon}\|_{L^{2n/(m-1)}(B_{r_1})} \le C' \frac{(1+\upsilon)^{\frac{q}{2(q-m)}}}{r_2 - r_1} \|w^{-\upsilon}\|_{L^2(B_{r_2})}.$$

Let $\Psi(\upsilon, r) = \left[\int_{B(r)} w^{-\upsilon}\right]^{\frac{1}{\upsilon}}$ and $\kappa = \frac{m}{n-1}$. Then (after replacing 2υ by υ)

$$\Psi(\kappa\upsilon, r_1) \le \left[C' \frac{(1+\upsilon)^{\frac{q}{2(q-m)}}}{r_2 - r_1}\right]^{\frac{1}{\upsilon}} \Psi(\upsilon, r_2).$$

Take any $p > 1$. Choose $\upsilon = \kappa^\mu p$ and $r_2 = 1 + \frac{1}{2^\mu}$ and $r_1 = 1 + \frac{1}{2^{\mu+1}}$. Then

$$\Psi(\kappa^{\mu+1}p, 1 + \frac{1}{2^{\mu+1}}) \le \left[C' \frac{(1+\kappa^\mu p)^{\frac{q}{2(q-m)}}}{\frac{1}{2^{\mu+1}}}\right]^{\frac{1}{\kappa^\mu p}} \Psi(\kappa^\mu p, 1 + \frac{1}{2^\mu}).$$

We have $\left[C' \frac{(1+\kappa^\mu p)^{\frac{q}{2(q-m)}}}{\frac{1}{2^{\mu+1}}}\right]^{\frac{1}{\kappa^\mu p}} \le C_*^{\frac{\mu}{\kappa^\mu}}$. Hence

$$\Psi(\kappa^{\mu+1}p, 1 + \frac{1}{2^{\mu+1}}) \le C_*^{\Sigma_{\upsilon=0}^{\mu} \frac{\upsilon}{\kappa^\upsilon}} \Psi(p, 2).$$

Letting $\mu \to \infty$, we get

$$\sup_{B_1} w^{-1} \le C_\# \Psi(p, 2).$$

(5.3) Now we want to show that for some $p > 0$ we have $\Psi(p,2) \le C_0\Psi(-p,2)$. Mulitplying $g^{\alpha\bar{\beta}}\partial_\alpha\partial_{\bar{\beta}} w \le \theta$ by $\frac{\eta^2}{w}$ and integrating with respect to the volume form of $g_{\alpha\bar{\beta}}$, we get

$$\int \eta^2 g^{\alpha\bar{\beta}}(\partial_\alpha \log w)(\partial_{\bar{\beta}} \log w) \le \int \theta \frac{\eta^2}{w} + \int 2\eta\, g^{\alpha\bar{\beta}}(\partial_\alpha \log w)(\partial_{\bar{\beta}}\, \eta)$$
$$\le \int \eta^2 + 2\int g^{\alpha\bar{\beta}}(\partial_\alpha\, \eta)(\partial_{\bar{\beta}}\, \eta) + \frac{1}{2}\int \eta^2 g^{\alpha\bar{\beta}}(\partial_\alpha \log w)(\partial_{\bar{\beta}} \log w).$$

Thus

$$\int \eta^2 g^{\alpha\bar{\beta}}(\partial_\alpha \log w)(\partial_{\bar{\beta}} \log w) \le 2\int \eta^2 + 4\int g^{\alpha\bar{\beta}}(\partial_\alpha\, \eta)(\partial_{\bar{\beta}}\, \eta).$$

We have

$$\|D \log w\|_{L^1(B_r)} \le C\, r^m\, \|D \log w\|_{L^2(B_r)} \le C'\, r^{2m-1}.$$

Now we use the following theorem of John-Nirenberg which we will later prove. If Ω is a convex subset of $\mathbb{R}^n$ and there exists a constant K such that $\|Df\|_{L^1(\Omega \cap B_r(x))} \le K\, r^{n-1}$ for a ball $B_r(x)$ of radius r centered at any point x, then there exist positive numbers σ_0 and C depending only on n such that $\int_\Omega \exp\left[\frac{\sigma}{K}\left|f - f_\Omega\right|\right] \le C(\operatorname{diam}\Omega)^n$, where $\sigma = \sigma_0(\operatorname{Vol}\Omega)(\operatorname{diam}\Omega)^{-n}$ and f_Ω is the average of f over Ω. In particular, $\left[\int_\Omega \exp(\frac{\sigma}{K} f)\right]\left[\int_\Omega \exp(-\frac{\sigma}{K} f)\right] \le C^2(\operatorname{diam}\Omega)^{2n}$.

We now apply this to $f = \log w$. Let $n = 2m$ and $\frac{\sigma}{C'} = p$. Then $\left[\int_{B_2} w^p\right]\left[\int_{B_2} w^{-p}\right] \le \hat{C}$ and $\Psi(p,2) \le C_0\Psi(-p,2)$. So $\sup_{B_1} w^{-1} \le C_\# C_0\Psi(-p,2)$ and $\|w\|_{L^p(B_2)} \le C_\#^{-1}C_0^{-1}\inf_{B_1} w$. Hence $\|u\|_{L^p(B_2)} \le C\left[\inf_{B_1} u + \|v\|_q\right]$. A change of scale yields

$$R^{-2m/p}\|v\|_{L^p(B_{2R})} \le C\left[\inf_{B_R} v + R^{\frac{2(q-m)}{q}} \|\theta\|_{L^q(B_{2R})}\right].$$

This concludes the derivation of the Harnack inequality.

(5.4) Now we prove the theorem of John-Nirenberg. Fix x in Ω. For y in Ω we apply the fundamental theorem of calculus to the restriction of f to the line segment joining x to y and then we average over y. We get

$$|f(x) - f_\Omega| \le \frac{d^n}{nV}\int_\Omega |x - y|^{1-n}\, |Df(y)|dy,$$

where d is the diameter of Ω and V is the volume of Ω. Now

$$\int_\Omega |x - y|^{1-n}\, |Df(y)|dy = \int_\Omega |x - y|^{(\frac{1}{q} - n)\frac{1}{q}}\, |x - y|^{(1 + \frac{1}{q} - n)(1 - \frac{1}{q})}\, |Df(y)|dy$$

$$\le \left[\int_\Omega |x - y|^{(\frac{1}{q} - n)}\, |Df(y)|dy\right]^{\frac{1}{q}} \left[\int_\Omega |x - y|^{(1 + \frac{1}{q} - n)}\, |Df(y)|dy\right]^{1 - \frac{1}{q}}.$$

Let $\nu(r) = \int_{B_r(x)\cap\Omega} |Df|$. Then

$$\int_\Omega |x - y|^{(1 + \frac{1}{q} - n)}\, |Df(y)|dy \le \int_0^d \rho^{(1 + \frac{1}{q} - n)}\, \nu'(\rho)d\rho$$

$$= d^{(1 + \frac{1}{q} - n)}\nu(d) + \int_0^d (1 + \frac{1}{q} - n)\, \rho^{(\frac{1}{q} - n)}\, \nu(\rho)d\rho$$

$$\le (1 + q(1 + \frac{1}{q} - n))\, K\, d^{\frac{1}{q}},$$

because $\nu(r) \le K r^{n-1}$. So

$$|f(x) - f_\Omega|^q$$
$$\leq \left(\frac{d^n}{nV}\right)^q \left[\int_\Omega |x - y|^{(\frac{1}{q} - n)} |Df(y)|dy\right](1 + q(1 + \frac{1}{q} - n))^{q-1} K^{q-1} d^{1 - \frac{1}{q}}.$$

Since

$$\int_\Omega \left[\int_\Omega |x - y|^{(\frac{1}{q} - n)} |Df(y)|dy\right]dx \leq \left[\sup_{y\in\Omega}\int_\Omega |x - y|^{(\frac{1}{q} - n)} dx\right]\nu(d)$$

$$\leq \omega_{n-1}\nu(d)\int_{\rho=0}^{d} \rho^{(\frac{1}{q} - 1)} d\rho = q\, \omega_{n-1}\nu(d)\, d^{\frac{1}{q}},$$

where ω_{n-1} is the volume of the unit (n-1)-sphere. Hence

$$\int_\Omega |f(x) - f_\Omega|^q dx \leq \left(\frac{d^n}{nV}\right)^q q\, \omega_{n-1}\nu(d)\, d^{\frac{1}{q}}(1 + q(1 + \frac{1}{q} - n))^{q-1} K^{q-1} d^{1 - \frac{1}{q}}$$
$$\leq \left(\frac{d^n}{nV}\right)^q q\, d^n \omega_{n-1}(1 + q(1 + \frac{1}{q} - n))^{q-1} K^q \leq \left(\frac{K}{\sigma}\right)^q C\, q^q d^n,$$

where $\sigma = \frac{nV}{d^n}$. So finally

$$\int_\Omega \exp\left[\frac{\sigma}{K} |f(x) - f_\Omega|\right] \leq C' d^n.$$

§6. *Historical Note.*

The existence of Kähler-Einstein metrics for the case of negative and zero anticanonical class is the work of Calabi, Aubin, and Yau. For the case of zero anticanonical class Calabi formulated the problem in the form of the Monge-Ampère equation [C1,C2,C3]. He proved the uniqueness of the solution [C3, pp.86-87] and laid out the program of proving the existence by the method of continuity and pointed out the openness and the need of *a priori* estimates [C3, pp.87-89]. Orginally the third order *a priori* estimate of φ was used instead of the *a priori* estimate for the Hölder norm of the second order derivative of φ. For the third-order estimate Aubin used the inequality $|\Delta' S - T| \le C(S + \sqrt{S})$, where

$$S = g'^{\alpha\bar\beta} g'^{\gamma\bar\delta} g'^{\lambda\bar\mu}\, \varphi_{\alpha\bar\delta\lambda}\, \varphi_{\bar\beta\gamma\bar\mu},$$

$$T = g'^{\alpha\bar\beta} g'^{\gamma\bar\delta} g'^{\lambda\bar\mu} g'^{\sigma\bar\tau} (\varphi_{\bar\beta\gamma\bar\mu\sigma}\, \varphi_{\alpha\bar\delta\lambda\bar\tau} + \varphi_{\alpha\gamma\bar\mu\sigma}\, \varphi_{\bar\beta\bar\delta\lambda\bar\tau})$$

and the subscripts of φ denote covariant differentiation with respect to $g_{\alpha\bar\beta}$ [A1,p.412,(5h)]. For the third-order estimate Yau used the inequality $|\Delta' S - T'| \le CS + C'$ of Calabi [Y2, pp.360-361 and Appendix A], where

$$T' = g'^{\alpha\bar\beta} g'^{\gamma\bar\delta} g'^{\lambda\bar\mu} g'^{\sigma\bar\tau} (\psi_{\bar\beta\gamma\bar\mu\sigma}\, \psi_{\alpha\bar\delta\lambda\bar\tau} + \theta_{\alpha\bar\mu\gamma\sigma}\, \theta_{\bar\beta\lambda\bar\delta\bar\tau})$$

with

$$\psi_{\bar\alpha\beta\bar\gamma\delta} = \varphi_{\bar\alpha\beta\bar\gamma\delta} - \varphi_{\bar\alpha\sigma\bar\gamma}\varphi_{\bar\tau\beta\delta} g'^{\sigma\bar\tau},$$

$$\theta_{\alpha\bar\beta\gamma\delta} = \varphi_{\alpha\bar\beta\gamma\delta} - (\varphi_{\bar\tau\alpha\delta}\varphi_{\sigma\bar\beta\gamma} + \varphi_{\bar\tau\alpha\gamma}\varphi_{\sigma\bar\beta\delta}) g'^{\sigma\bar\tau}.$$

Calabi did the real case of the above inequality for the third-order estimate [C4] and the transplantation to the complex case was done by Nirenberg [Y2,p.341]. Calabi's third-order estimate, being motivated by geometric considerations, is more natrual and easier to follow. The *a priori* estimate for the Hölder norm of the second derivative of φ given here is due to Evans [E1,E2] with simplification by Trudinger [T].

For the second order *a priori* estimate Aubin [A2, p.120, (β)] used the computation of $\Delta'(\log(m+\Delta\varphi) - k\varphi)$ and Yau [Y2,p.351,(2.22)] used $\Delta'(e^{-C\varphi}(1+\Delta\varphi))$. In our presentation here we use $\Delta'(\log(m+\Delta\varphi) - k\varphi)$, mainly because there is an analogous computation in the estimates for

Hermitian-Einstein metrics for stable bundles. Another insignificant reason is that the subharmonicity of a function contains more information than the subharmonicity of its exponential.

The very easy *a priori* zeroth order estimate for the case of negative anticanonical class was first given by Aubin [A2,p.120,(α)]. In that paper he gave his proof of the existence of Kähler-Einstein metrics for the negative anticanonical class case [A2,p.120, Th.2]. He did not apply his *a priori* estimates to the continuity method. Instead he used the method of minimizing the global L^2 norm of $\Delta^p\left(\log\left(\frac{\det(g_{i\bar{j}}+\varphi_{i\bar{j}})}{\det(g_{i\bar{j}})}\right)-\varphi-F\right)$. The method of continuity is easier.

Yau made the important contribution of using Moser's method of integration by parts and iteration by Sobolev inequalities, which Moser devised for Moser's proof of the Harnack inequality [Mo2], to get the *a priori* zeroth order estimate for the surface case when the anticanonical class is zero [Y2, pp.352-356]. In that paper he also gave a different and rather involved argument to handle the higher dimensional *a priori* zeroth order estimate for the case of zero anticanonical class [Y2, pp.352-359] whereby he obtained the existence of Kähler-Einstein metrics for the zero anticanonical class case. After learning Yau's *a priori* zeroth order estimate for the surface case through J. Kazdan [K], Aubin wrote up a proof of the higher dimensional *a priori* zeroth order estimate for the zero anticanonical class case using Moser's method of integration by parts and iteration by Sobolev inequalities and he presented his proof in the language of determinants [A3, pp.91-94]. Bourguignon independently obtained essentially the same proof as Aubin's and he used the language of exterior algebra instead of the language of determinants [Y2, pp.410-411]. Bourguignon's proof was given in an exposition of Averous and Deschamps in Séminaire Palaiseau published in the Astérisque [A-D]. The proof given here is taken from the exposition of Averous and Deschamps.

CHAPTER 3. UNIQUENESS OF KÄHLER-EINSTEIN METRICS UP TO BIHOLOMORPHISMS

In this chapter we prove the following theorem due to Bando-Mabuchi [B-M].

Theorem. Suppose M is a compact Kähler manifold of complex dimension m with positive anticanonical line bundle. Suppose there are two Kähler-Einstein metrics in the anticanonical class. Then some element in the connected component of the identity in the biholomorphism group of M maps one of the Kähler-Einstein metrics to the other.

§1. *The Role of Holomorphic Vector Fields*

Before we prove the theorem, we would like to give first a general discussion on Kähler-Einstein metrics on compact Kähler manifolds with positive anticanonical class to explain why Kähler-Einstein metrics are expected to be unique up to the action of holomorphic vector fields. The difference between the unsolved case of positive anticanonical class and the solved cases of negative and zero anticanonical class is the zeroth order estimates. The zeroth order estimate for the case of negative anticanonical class is immediate from the maximum principle. The zeroth order estimate for the case of zero anticanonical class is a consquence of integration by parts and the Moser iteration technique. For the case of positive anticanonical class there is no zeroth order estimate in general, because of obstructions to the existence of Kähler-Einstein metrics. However, for the case of positive anticanonical class, because the Ricci curvature is positive, there is a lower bound on the Green's function and one can reduce the supremum norm estimate of φ to some integral norm estimate of φ. Let us now look at the lower bound of the Green's function.

(1.1) Let (M,g) be a compact Riemannian manifold of real dimension n. Let $(n-1)K$ be a lower bound of the Ricci curvature of M and let $V = V_g$ be its volume and $D = D_g$ be its diameter. Let $K_0 = K\,D_g^2$. Then there exists a positive number $\gamma = \gamma(n,K_0)$ depending only on n and K_0 such that the Green's function $G_g(x,y)$ for the Riemannian metric g is bounded from below by $-\gamma \frac{D_g^2}{V_g}$. A discussion of this statement is given in Appendix A of this Chapter. Here the Green's function is defined by the following identity

$$f(x) = \frac{1}{V_g}\int_M f + \int_M G_g(x,y)(-\Delta_g f),$$

where Δ_g is the (negative) Laplacian for the Riemannian metric g. If $-A$ is the lower bound of G_g and $\Delta_g f \geq -\kappa$, then

$$\begin{aligned} f(x) &= \frac{1}{V_g}\int_M f + \int_M G_g(x,y)(-\Delta_g f) \\ &= \frac{1}{V_g}\int_M f + \int_M (G_g(x,y) + A)(-\Delta_g f) \\ &\leq \frac{1}{V_g}\int_M f + \int_M (G_g(x,y) + A)\ \kappa \\ &= \frac{1}{V_g}\int_M f + \kappa A. \end{aligned}$$

For the problem of Kähler-Einstein metrics for a compact Kähler manifold M of positive anti-canonical line bundle we consider the solution of the Monge-Ampère equation

$$\frac{\det(g_{i\bar{j}} + \partial_i\partial_{\bar{j}}\varphi)}{\det(g_{i\bar{j}})} = e^{-t\varphi+F}$$

by the continuity method for $0 \leq t \leq 1$, where F satisfies $R_{i\bar{j}} - g_{i\bar{j}} = \partial_i\partial_{\bar{j}}F$ with $\int_M e^F = \mathrm{Vol}\ M$ and $R_{i\bar{j}}$ is the Ricci curvature of the Kähler metric $g_{i\bar{j}}$. This Monge-Ampère equation is formulated in order to yield the equation $R'_{i\bar{j}} = tg'_{i\bar{j}} + (1-t)g_{i\bar{j}}$ for the Ricci curvature $R'_{i\bar{j}}$ of the Kähler metric $g'_{i\bar{j}} = g_{i\bar{j}} + \partial_i\partial_{\bar{j}}\varphi$ so that $R'_{i\bar{j}}$ admits a positive lower bound. In this case $\Delta\varphi > -m$ and $\Delta'\varphi = \Sigma_j \frac{\varphi_{j\bar{j}}}{1+\varphi_{j\bar{j}}} = m \quad \Sigma_j \frac{1}{1+\varphi_{j\bar{j}}} < m$. The volume of M with respect to $g'_{i\bar{j}}$ equals to the volume of M with respect to $g_{i\bar{j}}$. Let A' be the lower bound of the Green's function $G_{g'}$ of the negative Laplacian of the Kähler metric $g'_{i\bar{j}} = g_{i\bar{j}} + \partial_i\partial_{\bar{j}}\varphi$. Then

$$\sup_M \varphi \leq \frac{1}{V_g}\int_M \varphi \ + mA$$

$$\sup_M(-\varphi) \leq \frac{-1}{V_g}\int_M \varphi \, dV' \ + mA'.$$

When the Ricci curvature is bounded from below by a positive constant $(2m-1)a^2$, the diameter of the manifold is bounded from above by $\frac{\pi}{a}$ by the theorem of Bonnet-Myers (see *e.g.* [B-C, p. 256, Corollary 2]). Since the Ricci curvature $R'_{i\bar{j}}$ of the Kähler metric $g'_{i\bar{j}}$ satisfies $R'_{i\bar{j}} = tg'_{i\bar{j}} + (1-t)g_{i\bar{j}}$ and is bounded from below by t, it follows that for $0 < t \leq 1$ the oscillation of φ (which is defined as $\sup_M\varphi - \inf_M\varphi$) is bounded by a constant times $\frac{1}{t}$ if one has an upper bound of $\int_M \varphi \, dV - \varphi \, dV'$. A bound on the oscillation of φ gives us the zeroth order estimate of φ because of the normalization that $\int_M e^{-t\varphi + F} = \mathrm{Vol}\, M$. Let us denote this integral $\int_M \varphi \, dV - \varphi \, dV'$ by I. For the continuity method we want it to be bounded from above as t approaches some t_* from below. We now introduce a function which is equivalent to I and which has the property that it is a nondecreasing function of t. Let us now look at this equivalent function. We introduce it in the following general setting.

(1.2) Let M be a compact Kähler manifold of complex dimension m with a Kähler form ω_0 in its anticanonical class. We denote by F_{ω_0} the function with $\mathrm{Ricci}(\omega_0) - \omega_0 = \sqrt{-1}\partial\bar{\partial}F_{\omega_0}$ with $\int_M \exp F_{\omega_0} = \mathrm{Vol}\, M$.

Let φ be a smooth real-valued function on M. Let $\omega_\varphi = \omega_0 + \sqrt{-1}\partial\bar{\partial}\varphi$. Define $I = \int\varphi(\omega_0^m - \omega_\varphi^m)$ and $J = \int_{s=0}^1 \left[\int\varphi(\omega_0^m - \omega_{s\varphi}^m)\right]ds$. To emphasize the dependence of I and J on ω_φ and ω_0 we also write I and J as $I(\omega_\varphi, \omega_0)$ and $J(\omega_\varphi, \omega_0)$ respectively. The function that is equivalent to I is I - J. The equivalence is a matter of simple algebra. Let us first get this equivalence.

$$\begin{aligned}
J &= \int_{s=0}^1 \left[\int\varphi(\omega_0^m - \omega_{s\varphi}^m)\right]ds \\
&= -\int_{s=0}^1 \left[\Sigma_{j=0}^{m-1}\int\varphi\binom{m}{j+1}(\sqrt{-1}\partial\bar{\partial}\varphi)^{j+1}\omega_0^{m-j-1}\right]ds \\
&= \int_{s=0}^1 \left[\Sigma_{j=0}^{m-1}\int(\sqrt{-1}\partial\varphi\wedge\bar{\partial}\varphi)\binom{m}{j+1}(\sqrt{-1}\partial\bar{\partial}\varphi)^j\omega_0^{m-j-1}\right]ds \\
&= \Sigma_{j=0}^{m-1}\int(\sqrt{-1}\partial\varphi\wedge\bar{\partial}\varphi)\frac{1}{j+2}\binom{m}{j+1}(\sqrt{-1}\partial\bar{\partial}\varphi)^j\omega_0^{m-j-1} \\
&= \Sigma_{j=0}^{m-1}\int(\sqrt{-1}\partial\varphi\wedge\bar{\partial}\varphi)\frac{1}{j+2}\binom{m}{j+1}\Sigma_{k=0}^{j}(-1)^{j-k}\binom{j}{k}\omega_\varphi^k\omega_0^{j-k}\omega_0^{m-j-1} \\
&= \Sigma_{k=0}^{m-1}\int(\sqrt{-1}\partial\varphi\wedge\bar{\partial}\varphi)\{\Sigma_{j=k}^{m-1}(-1)^{j-k}\binom{j}{k}\binom{m}{j+1}\frac{1}{j+2}\}\omega_0^{m-k-1}\omega_\varphi^k \\
&= \Sigma_{k=0}^{m-1}\int(\sqrt{-1}\partial\varphi\wedge\bar{\partial}\varphi)\left(1 - \frac{k+1}{m+1}\right)\omega_0^{m-k-1}\omega_\varphi^k,
\end{aligned}$$

because

$$\begin{aligned}
\Sigma_{j=k}^{m-1}(-1)^{j-k}\binom{j}{k}\binom{m}{j+1}\frac{1}{j+2} &= \Sigma_{j=k}^{m-1}(-1)^{j-k}\binom{j}{k}\binom{m-1}{j}\frac{m}{(j+1)(j+2)} \\
&= \Sigma_{j=k}^{m-1}(-1)^{j-k}\binom{m-1}{k}\binom{m-k-1}{m-j-1}\frac{m}{(j+1)(j+2)} \\
&= \Sigma_{j=k}^{m-1}(-1)^{j-k}\binom{m-1}{k}\binom{m-k-1}{j-k}\frac{m}{(j+1)(j+2)} \\
&= \Sigma_{j=0}^{m-k-1}(-1)^{j}\binom{m-1}{k}\binom{m-k-1}{j}\frac{m}{(j+k+1)(j+k+2)} \\
&= \Sigma_{j=0}^{m-k-1}(-1)^{j}\binom{m-1}{k}\binom{m-k-1}{j}m\left[\frac{1}{j+k+1} - \frac{1}{j+k+2}\right] \\
&= m\binom{m-1}{k}\left[\frac{1}{k+1}\binom{m}{k+1}^{-1} - \frac{1}{k+2}\binom{m+1}{k+2}^{-1}\right] \\
&\qquad \text{(see the binomial coefficient identity below)} \\
&= m\binom{m-1}{k}\left[\frac{1}{m}\binom{m-1}{k}^{-1} - \frac{(k+1)}{(m+1)m}\binom{m-1}{k}^{-1}\right] \\
&= 1 - \frac{k+1}{m+1}\quad .
\end{aligned}$$

Comparing this with

$$\begin{aligned} I = \int\varphi(\omega_0^m - \omega_\varphi^m) &= \int\varphi(\omega_0 - \omega_\varphi)\Sigma_{k=0}^{m-1}\omega_0^{m-k-1}\omega^k \\ &= \Sigma_{k=0}^{m-1}\int\varphi(-\sqrt{-1}\partial\bar{\partial}\varphi)\omega_0^{m-k-1}\omega^k \\ &= \Sigma_{k=0}^{m-1}\int(\sqrt{-1}\partial\varphi\wedge\bar{\partial}\varphi)\omega_0^{m-k-1}\omega^k, \end{aligned}$$

we get

$$\frac{1}{m+1}I \le J \le (1 - \frac{1}{m+1})I$$

and

$$\frac{1}{m+1}I \le I-J \le \frac{m}{m+1} I.$$

This shows that the function I and $I - J$ are equivalent. The functions I and J with their relations were first introduced by Aubin [A5]. The function $I - J$ could be very loosely regarded as some form of an analog of the Donaldson functional discussed in §5 of Chapter 1. The above verification of the inequality involving I and J by binomial coefficients was done for me by Alan Fekete.

We now establish the following identity for binormial coefficients $\Sigma_{j=0}^{m-k-1}(-1)^j\binom{m-k-1}{j}\frac{1}{j+k+1} = \frac{1}{k+1}\binom{m}{k+1}^{-1}$ that was used above. On the one hand, we have

$$(1-x)^{m-k-1} = \Sigma_{j=0}^{m-k-1}(-1)^j\binom{m-k-1}{j}x^j.$$

$$(1-x)^{m-k-1}x^k = \Sigma_{j=0}^{m-k-1}(-1)^j\binom{m-k-1}{j}x^{j+k}.$$

$$\begin{aligned} \int_0^1(1-x)^{m-k-1}x^k dx &= \Sigma_{j=0}^{m-k-1}(-1)^j\binom{m-k-1}{j}\int_0^1 x^{j+k}dx \\ &= \Sigma_{j=0}^{m-k-1}(-1)^j\binom{m-k-1}{j}\frac{1}{j+k+1}. \end{aligned}$$

On the other hand, by integration by parts

$$\int_0^1 (1-x)^{m-k-1}x^k dx = k!\int_0^1 \frac{1}{(m-k)(m-k+1)\cdots(m-1)}(1-x)^{m-1}dx$$
$$= \frac{k!}{(m-k)(m-k+1)\cdots(m-1)m} = \frac{1}{k+1}\binom{m}{k+1}^{-1}.$$

Combining these two together we get our identity on binormial coefficients.

(1.3) Suppose φ is a function of some real parameter t. We want to find $\frac{d}{dt}J(\varphi)$. We have

$$\frac{d}{dt}J(\varphi) = \frac{d}{dt}\int_{s=0}^1 \left[\int_M \varphi(\omega_0^m - (\omega_0 + s\sqrt{-1}\partial\bar\partial\varphi)^m)\right]ds$$
$$= \int_{s=0}^1 \left[\int_M \dot\varphi(\omega_0^m - (\omega_0 + s\sqrt{-1}\partial\bar\partial\varphi)^m)\right]ds$$
$$+ \int_{s=0}^1 \left[\int_M \varphi(-m(\omega_0 + s\sqrt{-1}\partial\bar\partial\varphi)^{m-1}s\sqrt{-1}\partial\bar\partial\dot\varphi)\right]ds.$$

Now we integrate the first term $\int_{s=0}^1 \left[\int_M \dot\varphi(\omega_0^m - (\omega_0 + s\sqrt{-1}\partial\bar\partial\varphi)^m)\right]ds$ by parts with respect to s and get

$$\int_{s=0}^1 \left[\int_M \dot\varphi(\omega_0^m - (\omega_0 + s\sqrt{-1}\partial\bar\partial\varphi)^m)\right]ds$$
$$= \int_M s\dot\varphi(\omega_0^m - (\omega_0 + s\sqrt{-1}\partial\bar\partial\varphi)^m)\Big|_{s=0}^{s=1} + \int_{s=0}^1 \left[\int_M sm\dot\varphi(\omega_0 + s\sqrt{-1}\partial\bar\partial\varphi)^{m-1}\partial\bar\partial\varphi\right]ds$$
$$= \int_M \dot\varphi(\omega_0^m - (\omega_0 + \sqrt{-1}\partial\bar\partial\varphi)^m) + \int_{s=0}^1 \left[\int_M sm\varphi(\omega_0 + s\sqrt{-1}\partial\bar\partial\varphi)^{m-1}\sqrt{-1}\partial\bar\partial\dot\varphi\right]ds$$

by Stokes' theorem. Hence

$$\frac{d}{dt}J(\varphi) = \int_M \dot\varphi(\omega_0^m - (\omega_0 + \sqrt{-1}\partial\bar\partial\varphi)^m)$$

and

$$\frac{d}{dt}(I - J) = - \int_M \varphi \frac{\partial}{\partial t}((\omega_0 + \sqrt{-1}\partial\bar{\partial}\varphi)^m).$$

In the case when the 1-parameter family of φ satisfies the Monge-Ampère equation $(\omega_0 + \sqrt{-1}\partial\bar{\partial}\varphi)^m = \omega_0^m \exp(-t\varphi+F_{\omega_0})$, we have $\Delta'\dot{\varphi} = -\varphi - t\dot{\varphi}$ and

$$\frac{d}{dt}(I - J) = \int_M \varphi \, (-\Delta'\dot{\varphi}) \, dV'$$
$$= \int_M (-\Delta'\dot{\varphi} - t\dot{\varphi}) \, (-\Delta'\dot{\varphi}) \, dV' \geq 0,$$

because $-\Delta' \geq t$. Since $I - J$ is a nondecreasing function of t we cannot use it to solve the Monge-Ampère equation by the continuity method starting from $t = 0$ and ending up with $t = 1$. However, we can use it to solve the Monge-Ampère equation by the continuity method by going backward in t to start with $t = 1$ and ending up with $t = 0$. So far as the existence of Kähler-Einstein metric is concerned this is completely useless, but it is useful for the proof of the uniqueness of the Kähler-Einstein metric.

(1.4) Suppose we have two Kähler-Einstein metrics with Kähler forms $\omega_0 + \sqrt{-1}\partial\bar{\partial}\varphi^{(\nu)}$, $\nu = 1,2$, in the same Kähler class ω_0. Then both functions $\varphi^{(\nu)}$, $\nu = 1,2$, are solutions of the Monge-Ampère equation $(\omega_0 + \sqrt{-1}\partial\bar{\partial}\varphi)^m = e^{-t\varphi+F}\,\omega_0^m$ for $t = 1$. Recall that $\int_M e^F = \mathrm{Vol}\,M$. We now solve backward the Monge-Ampère equation by the continuity method. To accommodate openness at $t = 0$, we consider the following new Monge-Ampère equation

$$\frac{\det(g_{i\bar{j}} + \partial_i\partial_{\bar{j}}\varphi)}{\det(g_{i\bar{j}})} = \left[\frac{1}{\mathrm{Vol}\,M}\int_M e^{-t\varphi+F}\right]^{-1} e^{-t\varphi+F}$$

with $\int_M \varphi = 0$. This new Monge-Ampère equation is equivalent to the old Monge-Ampère equation

$$\frac{\det(g_{i\bar{j}} + \partial_i\partial_{\bar{j}}\varphi)}{\det(g_{i\bar{j}})} = e^{-t\varphi+F}$$

for $0 < t \leq 1$. If φ is a solution of the new equation, then the function $\varphi + \frac{1}{t}\log\left[\frac{1}{\mathrm{Vol}\ M}\int_M e^{-t\varphi+F}\right]$ satisfies the old equation. Conversely if φ satisfies the old equation, then $\varphi - \frac{1}{\mathrm{Vol}\ M}\int_M \varphi$ satisfies the new one. We will show that we have openness for the new equation always for $0 \leq t < 1$. Suppose we do get two 1-parameter families $\varphi^{(\nu)}(t)$, $\nu = 1,2$, $0 \leq t \leq 1$, satisfying the new equation so that $\varphi^{(\nu)}(1) = \varphi^{(\nu)}$. At $t = 0$ we have $(\omega_0 + \sqrt{-1}\partial\bar{\partial}\varphi^{(\nu)}(0))^m = e^F\omega_0^m$. By the uniqueness of this kind of Monge-Ampère equation we have $\varphi^{(1)}(0) = \varphi^{(2)}(0)$. Now because of openness for $0 \leq t < 1$, from the implicit function theorem we get $\varphi^{(1)}(t) = \varphi^{(2)}(t)$ for $0 \leq t < 1$. So we have $\varphi^{(1)} = \varphi^{(2)}$ when $t = 1$.

We now look at the problem of getting the 1-parameter family $\varphi^{(\nu)}(t)$. We do this by the continuity method for the new equation starting with $t = 1$ and ending with $t = 0$. We have no trouble with the closedness for $0 < t \leq 1$, because $I - J$ for solution of the new equation is nondecreasing and we have an *a priori* zeroth order estimate for the solution of the old equation and therefore an *a priori* zeroth order estimate for the solution of the new equation. We need closedness at $t = 0$ and openness at every t in the interval $[0,1]$ for the new equation. We going to show that the only possible trouble with openness is at $t = 1$. First let us show that due to the lower bound of the Green's function we have no trouble with closedness at $t = 0$ for the new equation.

Since the oscillation of the solution φ of the old equation is bounded by a constant times $\frac{1}{t}$ for $0 < t \leq 1$, we have an *a priori* bound on the oscillation of $t\varphi$. From the Monge-Ampère equation $(\omega_0 + \sqrt{-1}\partial\bar{\partial}\varphi)^m = e^{-t\varphi+F}\omega_0^m$ we have an *a priori* bound on $(\omega_0 + \sqrt{-1}\partial\bar{\partial}\varphi)^m$. From the zeroth order estimate in the case of the zero anticanonical class in §2 of Chapter 2 we have an *a priori* bound on the oscillation of φ. Thus we have an *a priori* zeroth order estimate for $\varphi - \frac{1}{\mathrm{Vol}\ M}\int_M \varphi$ which is the corresponding solution for the new equation. So we have closedness at $t = 0$ for the new equation.

(1.5) We now check openness of the new equation at $t = 0$. Let

$$\tilde{\Phi}(t,\varphi) = \log \frac{\det(g_{i\bar{j}} + \partial_i\partial_{\bar{j}}\varphi)}{\det(g_{i\bar{j}})} + \log\left[\frac{1}{\mathrm{Vol}\ M}\int_M e^{-t\varphi+F}\right] + t\varphi - F.$$

The new equation is $\tilde{\Phi}(t,\varphi) = 0$. At $t = 0$ the operator $\psi \to D_\varphi\tilde{\Phi}(t,\varphi)\cdot\psi$ is simply $\psi \to \Delta\psi$ which is invertible for ψ satisfying $\int_M \psi = 0$. So with the condition $\int_M \varphi = 0$ we have openness of the new equation at $t = 0$.

We now check for $0 < t < 1$ the openness of the old equation (and therefore also the equivalent new equation). We let $M(\varphi) = (\omega_0 + \sqrt{-1}\partial\bar{\partial}\varphi)^m/\omega_0^m$ and $\Phi(t,\varphi) = \log M(\varphi) + t\varphi - F$. The old Monge-Amprère equation is simply $\Phi(t,\varphi) = 0$. We have openness if the operator $\psi \to D_\varphi\Phi(t,\varphi)\cdot\psi$ is invertible. Now $D_\varphi\Phi(t,\varphi)\cdot\psi = \Delta_\varphi\cdot\psi + t\psi$. So we have openness if and only if $\Delta_\varphi + t$ is invertible. From the Monge-Amprère equation we have $R'_{i\bar{j}} = tg'_{i\bar{j}} + (1-t)g_{i\bar{j}}$. If $\Delta_\varphi f = -tf$ for some function f, then $\Delta_\varphi\bar{\partial}f = -t\bar{\partial}f$ and integration by parts yields

$$t\|\bar{\partial}f\|^2 = (-\Delta_\varphi\bar{\partial}f,\bar{\partial}f) = \|\bar{\nabla}_\varphi\bar{\partial}f\|^2 + (\mathrm{Ric}_\varphi\,\bar{\partial}f,\bar{\partial}f), \tag{1.5.1}$$

which, together with $\mathrm{Ric}_\varphi > t$ implies $\bar{\partial}f = 0$ and $f \equiv$ constant when $t < 1$, contradicting $\Delta_\varphi f = -tf$ when $t > 0$. Thus we have openness for $0 < t < 1$. Here $\bar{\nabla}_\varphi$ means the covariant differential operator in the $(0,1)$ direction with respect to the Kähler metric $g'_{i\bar{j}}$ and Ric_φ is the operator defined naturally by the Ricci curvature $R'_{i\bar{j}}$.

We now look at the openness of the old equation at $t = 1$ which is equivalent to the openness of the new equation at $t = 1$. At $t = 1$ (*i.e.* when the metric is Kähler-Einstein) f is an eigenfunction for the eigenvalue 1 of the positive Laplacian $-\Delta_\varphi$ if and only if $\bar{\nabla}_\varphi \bar{\partial} f = 0$ which means that the vector field $\uparrow\bar{\partial} f$ of type (1,0) obtained by raising the index of $\bar{\partial} f$ with respect to $g'_{i\bar{j}}$ is holomorphic. (Conversely any holomorphic vector field X on M is of the form $\uparrow\bar{\partial} f$, because the (0,1)-form $\downarrow X$ associated to X by lowering its index is $\bar{\partial}$-closed and by the vanishing theorem of Kodaira the positivity of the anticanonical line bundle implies that $\downarrow X$ is $\bar{\partial}$-exact.) From the equation (1.5.1) we conclude that $\Delta_\varphi + 1$ is invertible if and only if M admits no nonzero holomorphic vector field. So we have proved the following. If a compact Kähler manifold with positive first Chern class admits no holomorphic vector field, then there can exist on it at most one Kähler-Einstein metric in the anticanonical class.

§2. *Proof of Uniqueness.*

(2.1) Let us now look at the case when M admits nonzero holomorphic vector fields. Let G be the connected component of the complex Lie group of all biholomorphisms of M. We want to show that Kähler-Einstein metrics on M are unique up to the action of G. In other words, in the space of Kähler-Einstein metrics of M in the anticanonical class the group G has only one orbit. Our strategy is as follows. If we have two orbits $\mathbf{O}_\nu$ $(\nu = 1,2)$ of G. Then we can find a Kähler metric ω of M in its anitcanonical class and an element $\theta_\nu = \omega + \sqrt{-1}\partial\bar{\partial}\lambda_{\theta_\nu}$ of $\mathbf{O}_\nu$ so that both Monge-Ampère equations $(\omega + \sqrt{-1}\partial\bar{\partial}\varphi)^m/\omega^m = \exp(-t\varphi+F_\omega)$ have openness at $t = 1$ and $\varphi = \lambda_{\theta_\nu}$. As in the case of no nonzero holomorphic vector fields, this would imply that θ_1 and θ_2 are equal, contradicting that the two $\mathbf{O}_\nu$ orbits are distinct. The element θ_ν of $\mathbf{O}_\nu$ will be chosen as the point of $\mathbf{O}_\nu$ where the infimum of $I(\omega,\theta) - J(\omega,\theta)$ is achieved. The choice of the Kähler metric ω is not arbitrary and requires some work. We now present the details of this strategy.

(2.2) Fix a Kähler metric in the anticanonical class of M and denote its Kähler form by ω_0. Suppose $\mathbf{O}$ is an orbit of G in the space of all Kähler-Einstein metrics of M in the anticanonical class. Take a Kähler-Einstein metric in $\mathbf{O}$ and let θ be its Kähler form. There exists a function λ_θ such that $\theta = \omega_0 + \sqrt{-1}\partial\bar{\partial}\lambda_\theta$. We have trouble proving uniqueness by the backward continuity method because the openness may fail at $t = 1$ due to the obstruction from the kernel of $\Delta_\theta + 1$. This kernel is finite-dimensional. If we look at the orthogonal complement of this kernel, on the orthogonal complement we would certainly have the invertibility of the operator and then we examine what happens in the finite-dimensional subspace.

Let H_θ be the kernel of $\Delta_\theta + 1$ and let $H_\theta^\perp$ be its orthogonal complement with respect to θ. The tangent space of the orbit $\mathbf{O}$ at θ is naturally isomorphic to the space of all holomorphic vector fields on M. We know that every holomorphic vector field on M is of the form $\uparrow\bar{\partial}f$ for some $f \in H_\theta$. So the tangent space of the orbit $\mathbf{O}$ at θ is naturally isomorphic also to H_θ. This natural isomorphism can be described as follows. An element of $\mathbf{O}$ is a Kähler-Einstein metric $\omega_0 + \sqrt{-1}\partial\bar{\partial}u$ and $\Phi(1,u) = 0$, where $\Phi(\cdot,\cdot)$ is the function defined in (1.5). When we have a 1-parameter family of such functions $u(s)$ near $s = 0$ so that $u(0) = \lambda_\theta$, the derivative of $u(s)$ with respect to s at $s = 0$ belongs to H_θ. For $f \in H_\theta$ sufficiently small there exist $\eta(f)$ such that $\Phi(1,\lambda_\theta + f + \eta(f)) = 0$. This simply means that $f \to \lambda_\theta + f + \eta(f)$ is a local map from the tangent space of $\mathbf{O}$ at λ_θ to $\mathbf{O}$. The derivative of $\eta(f)$ with respect to f is zero at $f = 0$.

To get openness at $t = 1$ means to solve the equation $\Phi(t,\varphi(t)) = 0$ for $1 - \epsilon < t \le 1$ with $\epsilon > 0$ and $\varphi(1) = \lambda_\theta$. We replace the unknown $\varphi(t)$ by $\lambda_\theta + f + \eta(f) + \psi$ with f in H_θ and ψ in $H_\theta^\perp$. The reason for this replacement is that now at $t = 1$ the solution $\lambda_\theta + f + \eta(f) + \psi$ of $\Phi(1,\lambda_\theta + f + \eta(f) + \psi) = 0$ is given by $\psi = 0$ because of the definition of the map η. Intuitively this is the same as using a product coordinate system at λ_θ with one set of coordinates $f + \eta(f)$ along the orbit **O** and another set of coordinates ψ along $H_\theta^\perp$.

Let P be the orthogonal projection operator onto H_θ with respect to θ. Break up the equation $\Phi(t,\lambda_\theta + f + \eta(f) + \psi) = 0$ into two parts corresponding to the decomposition into H_θ and $H_\theta^\perp$. We have

$$P\Phi(t,\lambda_\theta + f + \eta(f) + \psi) = 0$$
$$\Psi(t,\lambda_\theta + f + \eta(f) + \psi) = 0,$$

where $\Psi(t,\lambda_\theta + f + \eta(f) + \psi) = (1 - P)\Phi(t,\lambda_\theta + f + \eta(f) + \psi)$. For fixed t and f we want to solve for ψ. This is possible near $t = 1$ because the operator $\psi' \to (D_\psi\Psi)\psi' = (\Delta_\theta + 1)\psi'$ is invertible on $H_\theta^\perp$ at $t = 1$ and $f = 0$. So we have a solution $\psi = \psi_{t,f}$ with $\psi_{1,0} = 0$. Now we pluck it into the finite set of linear equations $P\Phi(t,\lambda_\theta + f + \eta(f) + \psi) = 0$ and get $P\Phi(t,\lambda_\theta + f + \eta(f) + \psi_{t,f}) = 0$.

(2.3) To simplify notations let $\Phi_0(t,f) = P\Phi(t,\lambda_\theta + f + \eta(f) + \psi_{t,f})$. As we observed earlier, from the definition of the map η we know that $\Phi_0(1,f) = 0$ for all f with $\psi_{1,f} = 0$. To get openness of the backward continuity method at $t = 1$ we have to solve the finite set of linear equations $\Phi_0(t,f) = 0$ for f in terms of t for $1 - \epsilon < t \leq 1$ so that $f = 0$ at $t = 1$. We cannot use directly the implicit function theorem, because we know that $\Phi_0(1,f) = 0$ for all f and the derivative of $\Phi_0(1,f)$ in f must be identically zero and cannot be invertible. We try to use the second derivative instead. Let $\Phi_1(t,f) = \frac{1}{t-1}\Phi_0(t,f)$. Instead of the equation $\phi_0(t,f) = 0$ we consider the equation $\phi_1(t,f) = 0$. We have $\phi_1(1,0) = 0$. We try now to apply the implicit function theorem by computing the derivative of $\phi_1(1,f)$ with respect to f at $f = 0$. Let us postpone the computation and write down first the result.

$$\left[D_f\phi_1\Big|_{t=1,f=0}\right](f') = f' - P\,\langle\partial\bar{\partial}(D_t\psi_{t,f})_{t=1,f=0},\ \partial\bar{\partial}f'\rangle_\theta .$$

We want to show that for a good choice of θ this is invertible in order to get openness. How to choose θ in the orbit $\mathbf{O}$ to make this invertible as a function of f' is by no means clear. In the case when θ is a critical point for the restriction of the function $I(\omega_0,\cdot) - J(\omega_0,\cdot)$ to $\mathbf{O}$, the expression for the operator $\left[D_f\phi_1\Big|_{t=1,f=0}\right](f')$ becomes more manageable when we take its global inner product with another element f'' of H_θ over M with respect to the metric θ. We will do this computation later. The global inner product is given by

$$\left(\left[D_f\phi_1\Big|_{t=1,f=0}\right](f'),\ f''\right)_{L^2(M,\theta)} = \int_M (1 + \tfrac{1}{2}\,\Delta_\theta\lambda_\theta) f'f''\,\frac{\theta^m}{m!}.$$

The key step of the strategy is that this global inner product agrees with the Hessian of $I(\omega_0,\theta) - J(\omega_0,\theta)$ at θ evaluated at the elements f' and f'' of H_θ which can naturally be regarded as the tangent space of $\mathbf{O}$ at θ. It is not clear how to explain this coincidence geometrically because the proof is through rather involved direct computations. This is the unsatisfactory part of the proof. It would be much better if there is another more geometric argement. We will verify the computation of the Hessian of $I(\omega_0,\theta) - J(\omega_0,\theta)$ later.

(2.4) To simplify notations we use $\iota(\theta)$ to denote the function $I(\theta,\omega_0) - J(\theta,\omega_0)$. To emphasize the dependence of ι on ω_0 we also write $\iota(\theta)$ as $\iota(\omega_0,\theta)$. Let θ_0 be the point in $\mathbf{O}$ where the infimum of ι is achieved. The infimum of ι on $\mathbf{O}$ is achieved at some point θ_0 of $\mathbf{O}$, because each θ in $\mathbf{O}$ satisfies the Monge-Ampére equation $(\omega_0 + \sqrt{-1}\partial\bar{\partial}\lambda_\theta)^m/\omega_0^m = \exp(-\lambda_\theta + F_{\omega_0})$ and the bound on $\iota(\theta)$ gives us zeroth order *a priori* estimate of λ_θ and we have from the Monge-Ampére equation also *a priori* Hölder estimate of the second derivative of λ_θ. At the infimum point θ_0 of $\iota|\mathbf{O}$ the Hessian of ι at θ evaluated at f' and f" of H_θ is nonnegative. So the operator $D_f\Phi_1\Big|_{t=1,f=0}$ is semidefinite. We are going to show that one can slightly change ω_0 to make the Hessian of ι at θ_0 evaluated at $H_{\theta_0} \times H_{\theta_0}$ positive-definite.

Recall that earlier we computed

$$\frac{d}{dt}(I(\omega_0,\omega_\varphi) - J(\omega_0,\omega_\varphi)) = -\int_M \varphi\frac{\partial}{\partial t}((\omega_0 + \sqrt{-1}\partial\bar{\partial}\varphi)^m)$$

when we have a 1-parameter family of φ parametrized by t. Hence

$$(D_{\lambda_\theta}\iota(\theta))(f') = (\sqrt{-1})^m\int_M \lambda_\theta\ m\theta^{m-1}\wedge\sqrt{-1}\partial\bar{\partial}f'$$

and for $f' \in H_\theta$ we have $\Delta_\theta f' = -f'$ which can be rewritten as $m\sqrt{-1}\partial\bar{\partial}f'\wedge\theta^{m-1} = -f'\theta^m$. As a consequence

$$(D_{\lambda_\theta}\iota(\theta))(f') = -\int_M \lambda_\theta\ f'\ \theta^m$$

for $f' \in H_\theta$. So $\theta \in \mathbf{O}$ is a critical point of $\iota|\mathbf{O}$ if and only if λ_θ is perpendicular to H_θ with respect to the volume form of θ.

For $0 < \epsilon < 1$ let $\omega_\epsilon = (1-\epsilon)\omega_0 + \epsilon\theta_0 = \omega_0 + \sqrt{-1}\partial\bar{\partial}(\epsilon\lambda_{\theta_0})$. Consider now the function $\iota(\omega_\epsilon,\theta)$ instead of $\iota(\omega_0,\theta)$. The point θ_0 on $\mathbf{O}$ is still a critical point of $\iota(\omega_\epsilon,\cdot)|O$ because $\theta_0 = \omega_\epsilon + \sqrt{-1}\partial\bar\partial((1-\epsilon)\lambda_{\theta_0})$ and

$$D_{\lambda_\theta}\iota(\omega_\epsilon,\theta)\Big|_{\theta=\theta_0}(f') = -m\int_M (1-\epsilon)\lambda_{\theta_0}\ f'\ \theta_0^m = 0$$

for $f' \in H_{\theta_0}$. The Hessian of this new function $\iota(\omega_\epsilon,\cdot)$ at θ_0 now is given by

$$\begin{aligned}\mathrm{Hess}_{\lambda_\theta}\ \iota(\omega_\epsilon,\theta)\Big|_{\theta=\theta_0}(f',f') &= \int_M (1 + \tfrac{1}{2}\Delta_{\theta_0}((1-\epsilon)\lambda_{\theta_0})f'^2\ \frac{\theta^m}{m!} \\ &= (1-\epsilon)\left[\mathrm{Hess}_{\lambda_\theta}\ \iota(\omega_0,\theta)\Big|_{\theta=\theta_0}(f',f')\right] + \epsilon\int_M f'^2\ \frac{\theta^m}{m!}\end{aligned}$$

which is strictly positive-definite. The point θ_0 is a strict local minimum point for $\iota(\omega_\epsilon,\cdot)|O$. Intuitively this is comparable to the situation of considering the function which is the square of the distance from a point on some submanifold of the Euclidean space to the origin. When we move the origin closer to the submanifold along the perpendicular line from the origin to the submanifold, the Hessian of the square distance function at the point of the submanifold closest to the origin becomes more positive. So we conclude that for any $0 < \epsilon < 1$ the Monge-Ampère equation

$$(\omega_\epsilon + \sqrt{-1}\partial\bar\partial\varphi)^m/\omega_\epsilon^m = \exp(-t\ \varphi + F_{\omega_\epsilon})$$

has openness at $t = 1$ and $\varphi = (1-\epsilon)\lambda_{\theta_0}$.

(2.5) Now suppose we have another **O'** of the group **G** distinct from **O**. Take $0 < \epsilon < 1$. Let θ_0' be a point of **O'** where the function $\iota(\omega_\epsilon,\cdot)|O'$ achieves its minimum. Take $0 < \delta < 1$. Let $\omega_\delta' = (1-\delta)\omega_\epsilon + \delta\ \theta_0'$. Then the Hessian of the function $\iota(\omega_\delta',\cdot)|O'$ at its local minimum point θ_0' is positive definite. Let θ_δ be the point of $\mathbf{O}$ where $\iota(\omega'_\delta,\cdot)|\mathbf{O}$ achieves its minimum. Since ω'_δ approaches ω_ϵ as $\delta \to 0$ and θ_0 is a strict local minimum for $\iota(\omega_\epsilon,\cdot)|\mathbf{O}$, we can choose for every sufficiently small positive δ a point θ_δ of $\mathbf{O}$ so that θ_δ is a strict local minimum of the function $\iota(\omega'_\delta,\cdot)|\mathbf{O}$ and θ_δ approaches θ_0 as $\theta \to 0$. Since the Hessian of $\iota(\omega_\epsilon,\cdot)|\mathbf{O}$ at θ_0 is positive definite, it follows that for δ sufficiently small the Hessian of $\iota(\omega'_\delta,\cdot)|\mathbf{O}$ at θ_δ is positive definite. Thus the Monge-Ampère equation

$$(\omega'_\delta + \sqrt{-1}\partial\bar{\partial}\varphi)^m / \omega'^m_\delta = \exp(-t\,\varphi + F_{\omega'_\delta})$$

has openness at $t = 1$ and φ with $\omega'_\delta + \sqrt{-1}\partial\bar{\partial}\varphi = \theta'_0$ and also at $t = 1$ and φ with $\omega'_\delta + \sqrt{-1}\partial\bar{\partial}\varphi = \theta_\delta$. So we conclude that θ_δ and θ'_0 are equal, contradicting that the orbit $\mathbf{O}'$ is distinct from the orbit $\mathbf{O}$. There is only one orbit of the group G in the space of all Kähler-Einstein metrics of M in the anticanonical class. This concludes the proof of the theorem of Bando-Mabuchi.

§3. *Computation of the Differential.*

(3.1) We now do the first of the two deferred computations, namely the computation of the differential of Φ_1 with respect to f.

$$\begin{aligned}
D_f\Phi_1\Big|_{t=1,f=0}(f') &= D_f\frac{\partial}{\partial t}\Phi_0\Big|_{t=1,f=0}(f') \\
&= P\, D_f\frac{\partial}{\partial t}\Phi(t,\lambda_\theta + f + \eta(f) + \psi_{t,f}) \\
&= P\, D_f\Big[\Delta_{\lambda_\theta+f+\eta(f)+\psi_{t,f}}(D_t\psi_{t,f}) + \lambda_\theta + f + \eta(f) + D_t\psi_{t,f}\Big]_{t=1,f=0} \\
&= P\,\Big[\Delta_{\lambda_\theta}(D_f D_t\psi_{t,f}) - \langle\partial\bar{\partial}(D_t\psi_{t,f}),\partial\bar{\partial}f'\rangle + f' + D_f D_t\psi_{t,f}\Big]_{t=1,f=0} \\
&= f' - P\langle\partial\bar{\partial}(D_t\psi_{t,f})\Big|_{t=1,f=0},\partial\bar{\partial}f'\rangle_\theta
\end{aligned}$$

because $P\left[(\Delta_{\lambda_\theta}+1)(D_fD_t\psi_{t,f})\right]$ vanishes by definition of P. Now we take the inner product with f''.

$$\left[D_f\phi_1\Big|_{t=1,f=0}(f'),\ f''\right]_{L^2(M,\theta)} = \int_M\left[f'f'' - f''\langle\partial\bar\partial(D_t\psi_{t,f})\Big|_{t=1,f=0},\partial\bar\partial f'\rangle_\theta\right]\frac{\theta^m}{m!}.$$

We now need the following identity which we will later verify:

$$(3.1.1)\qquad -\int_M f'\langle\partial\bar\partial\psi,\partial\bar\partial f''\rangle_\theta\ \theta^m = \int_M(f'f'' - \langle\partial f',\partial f''\rangle_\theta)((\Delta_\theta+1)\psi)\ \theta^m$$

for any $f',f''\in H_\theta$ and any smooth function ψ on M. Using this identity, we get

$$\begin{aligned}
&\left(\left[D_f\phi_1\Big|_{t=1,f=0}\right](f'),\ f''\right)_{L^2(M,\theta)}\\
&= \int_M\left[f'f'' + (f'f'' - \langle\partial f',\partial f''\rangle_\theta)\left((\Delta_\theta+1)(D_t\psi_{t,f})\Big|_{t=1,f=0}\right)\right]\frac{\theta^m}{m!}\\
&= \int_M\left[f'f'' - (f'f'' - \langle\partial f',\partial f''\rangle_\theta)\lambda_\theta\right]\frac{\theta^m}{m!}\\
&= \int_M(f'f'' + \frac{1}{2}\lambda_\theta\Delta_\theta(f'f''))\ \frac{\theta^m}{m!}\\
&\qquad\qquad\qquad (\text{using } \Delta_\theta f' = -f' \text{ and } \Delta_\theta f'' = -f'')\\
&= \int_M(1 + \frac{1}{2}\Delta_\theta\lambda_\theta)f'f''\ \frac{\theta^m}{m!}.
\end{aligned}$$

Here we have used $(\Delta_\theta+1)(D_t\psi_{t,f})\Big|_{t=1,\varphi=0} = -\lambda_\theta$. This can be seen as follows. Since at $t = 1$ and $f = 0$ we have

$$\begin{aligned}
0 = \frac{\partial}{\partial t}\Psi(t,\lambda_\theta + f + \eta(f) + \psi_{t,f}) &= (1 - P)\frac{\partial}{\partial t}\Phi(t,\lambda_\theta + f + \eta(f) + \psi_{t,f})\\
&= (1 - P)\left[\Delta_{\lambda_\theta}(D_t\psi_{t,f}) + \lambda_\theta + D_t\psi_{t,f}\right]_{t=1,f=0}
\end{aligned}$$

and since $(1-P)(\Delta_{\lambda_\theta}+1) = \Delta_{\lambda_\theta}+1$, it follows that $(\Delta_{\lambda_\theta}+1)(D_t\psi_{t,f})\Big|_{t=1,f=0} = -(1-P)\lambda_\theta$ which is equal to $-\lambda_\theta$, because θ being critical for $\iota(\omega_0,\cdot)$ means that λ_θ belongs to $H_\theta^\perp$.

(3.2) Now we verify identity (3.1.1). Since f'' belongs to H_θ and θ is Kähler-Einstein, we know that $\uparrow\bar\partial f''$ is a holomorphic vector field and $f''^{\alpha}{}_{\bar\beta} = 0$. So we have

$$(3.2.1)\qquad \begin{aligned}\Delta_\theta\langle\partial\psi,\partial f''\rangle &= \Delta_\theta(\psi_\alpha f''^{\alpha}) = \psi_{\alpha\bar\beta}\, f''^{\alpha}{}_{\beta} + \psi_{\alpha\bar\beta\beta}\, f''^{\alpha}\\ &= \langle\partial\bar\partial\psi,\partial\bar\partial f''\rangle_\theta + \langle\partial(\Delta_\theta\psi),\partial f''\rangle_\theta\end{aligned}$$

Here the raising of indices is done with the Kähler metric θ and summation over repeated indices (either one in the subscript position and one in the subscript position of the same type or both in the subscript or superscript position of different types) are used. Moreover, the scripts denote covariant differentiation with respect to θ. Note that $\psi_{\alpha\bar\beta\beta} = \psi_{\bar\beta\alpha\beta}$ because of the torsion-free condition and $\psi_{\bar\beta\alpha\beta} = \psi_{\bar\beta\beta\alpha}$ because of the vanishing of the curvature tensor when two skew-symmetric indices are of the same type. Let $\xi = (\Delta_\theta + 1)\psi$. Then

$$\int_M (f'f'' - \langle\partial f',\partial f''\rangle_\theta)\xi\ \theta^m = -\int_M (f'\partial\bar\partial f'' + \partial f'\wedge\bar\partial f'')\xi\wedge m\theta^{m-1}$$

$$\text{(using } \Delta_\theta f'' = -f''\text{)}$$

$$= -\int_M \xi\ \partial(f'\bar\partial f'')\wedge m\theta^{m-1} = \int_M \partial\xi\wedge(f'\bar\partial f'')\wedge m\theta^{m-1}$$

$$= \int_M f'\langle\partial\xi,\partial f''\rangle\ \theta^m = \int_M f'\langle\partial(\Delta_\theta\psi),\partial f''\rangle\ \theta^m + \int_M f'\langle\partial\psi,\partial f''\rangle\ \theta^m$$

$$= \int_M f'\left[\Delta_\theta\langle\partial\psi,\partial f''\rangle_\theta - \langle\partial\bar\partial\psi,\partial\bar\partial f'\rangle_\theta\right]\theta^m + \int_M f'\langle\partial\psi,\partial f''\rangle\ \theta^m$$

$$\text{(by (3.2.1))}$$

$$= \int_M \left[(\Delta_\theta f')\langle\partial\psi,\partial f''\rangle_\theta - f'\langle\partial\bar\partial\psi,\partial\bar\partial f'\rangle_\theta\right]\theta^m + \int_M f'\langle\partial\psi,\partial f''\rangle\ \theta^m$$

$$= - \int_M f' \langle \partial\bar{\partial}\psi, \partial\bar{\partial} f' \rangle_\theta \, \theta^m$$

(using $(\Delta_\theta + 1)\varphi' = 0$).

§4. *Computation of the Hessian.*

(4.1) Now we compute the Hessian of $\iota(\omega_0, \cdot)|O$. We take $\theta = \theta_{s,t}$ parametrized by two real variables s and t. So $\lambda_{\theta_{s,t}}$ depend on the two real parameters s and t. For notational simplicity we write $\lambda_{\theta_{s,t}}$ simply as $\lambda_{s,t}$ and write $\Delta_{\theta_{s,t}}$ simply as $\Delta_{s,t}$. Let $f' = \frac{\partial}{\partial s} \lambda_{s,t}\Big|_{s=0,t=0}$ and $f'' = \frac{\partial}{\partial t} \lambda_{s,t}\Big|_{s=0,t=0}$. From

$$\frac{d}{dt}(I(\omega_0, \omega_\varphi) - J(\omega_0, \omega_\varphi)) = - \int_M \varphi \frac{\partial}{\partial t}((\omega_0 + \sqrt{-1}\partial\bar{\partial}\varphi)^m)$$

we have

$$\frac{\partial}{\partial t} \iota(\theta) = - \int_M \lambda_{s,t} \Delta_{s,t} (\frac{\partial}{\partial t} \lambda_{s,t}) \frac{\theta^m_{s,t}}{m!}.$$

Differentiating $\log M(\lambda_{s,t}) + \lambda_{s,t} - F_{\omega_0} = 0$ with respect to t, we get

$$\Delta_{s,t}(\frac{\partial}{\partial t}\lambda_{s,t}) + \frac{\partial}{\partial t}\lambda_{s,t} = 0. \tag{4.1.1}$$

Using the fact that the derivative $(A^{-1})'$ of the inverse A^{-1} of a nonsingular matrix A equals $-A^{-1}A'A^{-1}$ (where A' is the derivative of A), we get by differentiating (4.1.1) with respect to s

$$-\langle \partial\bar{\partial}(\frac{\partial}{\partial s}\lambda_{s,t}), \partial\bar{\partial}(\frac{\partial}{\partial s}\lambda_{s,t}) \rangle_{\theta_{s,t}} + (\Delta_{s,t} + 1)(\frac{\partial^2}{\partial s \partial t} \lambda_{s,t}) = 0.$$

So

$$(\Delta_{s,t}+1)\left(\frac{\partial^2}{\partial s\partial t}\lambda_{s,t}\right)\Big|_{s=0,t=0} = \langle\partial\bar{\partial}f',\partial\bar{\partial}f''\rangle_{\theta_{0,0}}$$

$$= \Delta_\theta\langle\partial f',\partial f''\rangle_\theta - \langle\partial(\Delta_\theta f'),\partial f''\rangle_\theta$$

(because $\uparrow\bar{\partial}f'$ is holomorphic)

$$= (\Delta_\theta+1)\langle\partial f',\partial f''\rangle_\theta \qquad \text{(because } \Delta_\theta f' = -f'\text{)}.$$

So $\left(\frac{\partial^2}{\partial s\partial t}\lambda_{s,t}\right)\Big|_{s=0,t=0}$ is equal to $\langle\partial f',\partial f''\rangle_\theta$ modulo H_θ.

(4.2) We now assume that $\theta_{0,0}$ is a critical point for the function $\iota(\omega_0,\cdot)|0$ so that $\lambda_{0,0}$ belongs to $H^\perp_{\theta_{0,0}}$. Using $\Delta_{s,t}(\frac{\partial}{\partial t}\lambda_{s,t}) = -\frac{\partial}{\partial t}\lambda_{s,t}$ derived above, we have

$$(\text{Hess }\iota)_\theta(f',f'') = -\frac{\partial}{\partial s}\int_M \lambda_{s,t}\Delta_{s,t}\left(\frac{\partial}{\partial t}\lambda_{s,t}\right)\frac{\theta^m_{s,t}}{m!}$$

$$= -\frac{\partial}{\partial s}\int_M \lambda_{s,t}\left(\frac{\partial}{\partial t}\lambda_{s,t}\right)\frac{\theta^m_{s,t}}{m!}$$

$$= \int_M \left[f'f'' + \lambda_\theta\left(\frac{\partial^2}{\partial s\partial t}\lambda_{s,t}\right)_{0,0} + \lambda_\theta f''(\Delta_\theta f')\right]\frac{\theta^m_{s,t}}{m!}$$

(last term from differentiating $\theta^m_{s,t}$)

$$= \int_M \left[f'f'' + \lambda_\theta\langle\partial f',\partial f''\rangle_\theta - \lambda_\theta f''f'\right]\frac{\theta^m_{s,t}}{m!}$$

(using $\lambda_\theta \in H^\perp_\theta$ and $\Delta_\theta f' = -f'$)

$$= \int_M \left[f'f'' + \frac{1}{2}\lambda_\theta\,\Delta_\theta(f'f'')\right]\frac{\theta^m_{s,t}}{m!}$$

$$= \int_M \left(1 + \frac{1}{2}\Delta_\theta\lambda_\theta\right)f'f''\,\frac{\theta^m_{s,t}}{m!}.$$

APPENDIX A. LOWER BOUNDS OF THE GREEN'S FUNCTION OF LAPLACIAN

In this appendix we discuss the result on the lower bound of the Green's function of a compact Riemannian manifold in terms of the diameter, the volume, and the (possibly negative) lower bound of the Ricci curvature. We will only deal with the case of real dimension at least three, because this is the only case we need for our application. Though the statements hold also for the case of real dimension two, due to the presence of a factor equal to the real dimension minus two in some intermediate steps the case of real dimension has to be treated separately.

(A.1) *Definition of Green's Function and the Heat Kernel.* Let (M,g) be a compact Riemannian manifold of real dimension n. Let $\square$ be the positive Laplacian of (M,g) for functions of M. First we introduce the Green's function. Let $\mathcal{H}_\nu$ be the Hilbert space of all functions on M whose derivatives up to and including order ν are L^2. Let $\|\cdot\|_\nu$ be the norm of the Hilbert space $\mathcal{H}_\nu$ and $(\cdot,\cdot)_\nu$ be its inner product. By integration by parts we have $((1+\square)f,f)_0 \geq C\|f\|_1^2$ for functions f, where C is a positive constant. Replace f by $(1+\square)^{-1}f$, we get

$$C\|(1+\square)^{-1}f\|_1^2 \leq (f,\ (1+\square)^{-1}f)_0 \leq \|f\|_0\|(1+\square)^{-1}f\|_0 \leq \|f\|_0\|(1+\square)^{-1}f\|_1$$

and $\|(1+\square)^{-1}f\|_1 \leq \frac{1}{\sqrt{C}}\|f\|_0$. This means that the map $(1+\square)^{-1}: \mathcal{H}_0 \to \mathcal{H}_0$ factors through the inclusion map $\mathcal{H}_1 \to \mathcal{H}_0$ and is therefore a compact map. So the eigenvalues of $(1+\square)^{-1}$ (which are clearly inside the interval $(0,1]$) form a discrete set with 0 as the only point of accumulation and all the eigenspaces of $(1+\square)^{-1}$ are finite-dimensional. The eigenvalues λ of $\square$ are related to the eigenvalues μ of $(1+\square)^{-1}$ by $\lambda = \frac{1}{1+\mu}$. So the eigenvalues of $\square$ (which are inside $[0,\infty)$) form a discrete set with ∞ as the only point of accumulation and all the eigenspaces of $\square$ are finite-dimensional. We take an orthonormal basis $\{f_i\}$ of $\mathcal{H}_0$ so that f_i is an eigenfunction for the eigenvalue λ_i with $\lambda_0 = 0$.

Let $H(x,y,t) = \Sigma_{i=0}^{\infty} e^{-\lambda_i t} f_i(x)f_i(y)$. Then $H(x,y,t)$ is the heat kernel in the sense that $(\frac{\partial}{\partial t} + \square_x) H(x,y,t) = 0$ and $\int_{y\in M} H(x,y,0)f(y)dy = f(x)$. So the function $f(x,t) = \int_{y\in M} H(x,y,t)f(y)dy$ satisfies the heat equation $(\frac{\partial}{\partial t} + \square) f(x,t) = 0$ and $f(x,0) = f(x)$. The heat kernel is everywhere positive. The reason is as follows. First $H(x,y,0)$ is positive everywhere, because $H(x,y,0)$ is reproducing and if it is negative on some open neighborhood $U \times V$ of (x_0,y_0), then it cannot be reproducing for a function which is positive and supported on V. Moreover, $H(x,y,t)$ satisfies the heat equation for any fixed value of y and the minimum of $H(x,y,t)$ in (x,t) must be achieved at $t = 0$ by the minimum principle.

Let $G(x,y) = \Sigma_{i=1}^{\infty} \frac{1}{\lambda_i} f_i(x)f_i(y)$. Then $G(x,y)$ is the Green's function. It has the property that

$$f(x) = \frac{1}{\text{Vol } M} \int_{y\in M} f(y)dy + \int_{y\in M} G(x,y)(\square_y f(y))dy,$$

as one can easily see by expanding $f(x) = \Sigma_{i=0}^{\infty} \alpha_i f_i(x)$ and using $f_0(x) \equiv \frac{1}{\sqrt{\text{Vol } M}}$. Since each f_i is orthogonal to the constant function f_0 for $i > 0$, it follows that $\int_{y\in M} G(x,y)dy \equiv 0$. The Green's function $G(x,y)$ is related to the heat kernel $H(x,y,t)$ in the following way. Let $G(x,y,t) = \Sigma_{i=1}^{\infty} e^{-\lambda_i t} f_i(x)f_i(y) = H(x,y,t) - \frac{1}{\text{Vol } M}$. Then $G(x,y) = \int_{t=0}^{\infty} G(x,y,t)dt$. We observe that since $H(x,y,t)$ is everywhere positive, it follows that $\int_{y\in M} |G(x,y,t)|dy \le 2$.

(A.2) *Lower Bound of Green's Function.* Suppose the Ricci curvature of the Riemannian manifold (M,g) of real dimension n is bounded from below by

$(n-1)K$ and $V = V_g$ be its volume, and $D = D_g$ be its diameter. Let $K_0 = K\, D_g^2$. The goal of this appendix is to prove that there exists a positive number $\gamma = \gamma(n, K_0)$ such that the Green's function $G_g(x,y)$ for the Riemannian metric g is bounded from below by $-\gamma \frac{D_g^2}{V_g}$.

We need the following Sobolev inequality. $\|df\|_{L^2} \geq C\|f\|_{L^{2n/(n-2)}}$ for $\int_M f = 0$, where $C = \kappa(n,\alpha)\frac{V_g^{1/n}}{D_g}$. This Sobolev inequality will be proved later. We want to apply this to the function $G(x,y,t)$ as a function of y. From the definition of $G(x,y,t)$ we have

$$\left(\frac{\partial}{\partial t} + \Box_x\right) G(x,y,t) = 0,$$

$$\int_{y\in M} G(x,y,t)dy = 0,$$

$$G(x,y,t+s) = \int_{y\in M} G(x,z,s)G(z,y,t)dy.$$

Differentiating both sides of the last equation with respect to t and setting $x = y$, we obtain

$$G'(x,x,t) = \int_{y\in M} G'(x,y,\tfrac{t}{2})G(x,y,\tfrac{t}{2})dy = \int_{y\in M} (-\Box_y G(x,y,\tfrac{t}{2}))G(x,y,\tfrac{t}{2})dy,$$

where $G'(x,x,t)$ means derivative with respect to t. Integrating by parts and using the above Sobolev inquality yield

$$-G'(x,x,t) = \int_{y\in M} |d_y G(x,y,\tfrac{t}{2})|^2 dy \geq C^2\left[\int_{y\in M} |G(x,y,\tfrac{t}{2})|^{\frac{2n}{n-2}}\right]^{\frac{n-2}{n}}.$$

By Hölder's inequality we have

$$\left[\int_{y\in M} |G(x,y,\tfrac{t}{2})|^2\right]^{\frac{n+2}{n}} \leq \left[\int_{y\in M} |G(x,y,\tfrac{t}{2})|^{\frac{2n}{n-2}}\right]^{\frac{n-2}{n}} \left[\int_{y\in M} |G(x,y,\tfrac{t}{2})|\right]^{\frac{4}{n}}.$$

It follows from $\int_{y\in M} |G(x,y,t)|\,dy \leq 2$ that

$$-G'(x,x,t) \geq C^2 \left[\int_{y\in M} |G(x,y,\tfrac{t}{2})|^2\right]^{\frac{n+2}{n}} = C^2 G(x,x,t)^{\frac{n+2}{n}}$$

after we absorb the factor $2^{-\frac{4}{n}}$ into C^2. Integrating the inequality

$$-G'(x,x,t)\, G(x,x,t)^{-\frac{n+2}{n}} \geq C^2$$

with respect to t and using

$$\lim_{t\to\infty} G(x,x,t) = \lim_{t\to\infty} \Sigma_{i=1}^{\infty} e^{-\lambda_i t} |f_i(x)|^2 \geq \Sigma_{i=1}^{\ell} |f_i(x)|^2 = \ell$$

for any finite ℓ, we obtain $\tfrac{n}{2} G(x,x,t)^{-\frac{2}{n}} \geq C^2 t$ and

$$G(x,x,t) \leq C^{-n}(\tfrac{n}{2})^{\frac{n}{2}} t^{-\frac{n}{2}} \quad \text{and}$$

$$|G(x,y,t)| \leq \sqrt{G(x,x,\tfrac{t}{2})}\,\sqrt{G(y,y,\tfrac{t}{2})} \leq C^{-n}(\tfrac{n}{2})^{\frac{n}{2}} t^{-\frac{n}{2}}.$$

Since $H(x,y,t)$ is positive, we have $G(x,y,t) \geq -\tfrac{1}{V_g}$. Now

$$G(x,y) = \int_{t=0}^{\infty} G(x,y,t)dt \geq -\int_{t=0}^{\tau} \frac{dt}{V_g} - \int_{t=\tau}^{\infty} C^{-n}(\tfrac{n}{2})^{\frac{n}{2}} t^{-\frac{n}{2}}dt$$

$$= -\frac{\tau}{V_g} - C^{-n}(\tfrac{n}{2})^{\frac{n}{2}} \tau^{-\frac{n-2}{2}}.$$

Set $\tau = D_g^2$. Then

$$G(x,y) \geq -\frac{D_g^2}{V_g} - \kappa(n,K_0)^{-n}\frac{D_g^n}{V_g}\left(\frac{n}{2}\right)^{\frac{n}{2}} D_g^{n-2} = -\gamma(n,K_0)\frac{D_g^2}{V_g}.$$

The above argument of obtaining the lower bound of the Green's function from the Sobolev inequality is due to Cheng and Li [C-L].

(A.3) *Reduction of Sobolev Inequality.* Before we prove the Sobolev inequality $\|df\|_{L^2} \geq C\|f\|_{L^{2n/(n-2)}}$ for $\int_M f = 0$ with $C = \kappa(n,\alpha)\frac{V_g^{1/n}}{D_g}$ that we used earlier, we first reduce it to the following form: $\|dg\|_{L^1} \geq C_0\|g\|_{L^{n/(n-1)}}$ when the volume of $\{g \leq 0\}$ equals to the volume of $\{g \geq 0\}$, where $C_0 \geq \kappa_0(n,K_0)\frac{V_g^{1/n}}{D_g}$. We use the convention that C_0 is the largest constant for which the inequality holds. We assume that the new inequality is known and derive the original inequality.

Choose a real number a such that the volume of $\{f \leq a\}$ equals the volume of $\{f \geq a\}$. Let $h = f - a$ and $g = (\operatorname{sgn} h)\,|h|^{2(n-1)/(n-2)}$, we have $|g|^{n/(n-1)} = |h|^{2n/(n-2)}$ and $dg = \frac{2(n-1)}{n-2}\,|h|^{n/(n-2)}\,dh$. Since the volume of $\{g \leq 0\}$ equals the volume of $\{g \geq 0\}$, from $\|dg\|_{L^1} \geq C_0\|g\|_{L^{n/(n-1)}}$ we have

$$C_0\left[\int_M |h|^{2n/(n-2)}\right]^{(n-1)/n} \leq \frac{n}{n-2}\int_M |h|^{n/(n-2)}|dh|$$
$$\leq \frac{n}{n-2}\left[\int_M |h|^{2n/(n-2)}\right]^{1/2}\left[\int_M |dh|^2\right]^{1/2}$$

and $\|dh\|_{L^2} \geq C_0\left(\frac{n-2}{n}\right)\|h\|_{L^{2n/(n-2)}}$. For any real number a and any $p > 1$, from integrating $a^2 \leq 2|f-a|^2 + 2|f|^2$ over M and Hölder's inequality it follows that

$$a^2 \operatorname{Vol} M \leq 2(\operatorname{Vol} M)^{2/n}\left[\int_M |f-a|^{2n/(n-2)}\right]^{(n-2)/n} + 2\int_M |f|^2$$

and

$$\text{(A.3.1)} \qquad |a| \leq \sqrt{2}\,(\operatorname{Vol} M)^{(2-n)/(2n)}\,\|f-a\|_{L^{2n/(n-2)}} + \sqrt{2}\,(\operatorname{Vol} M)^{-1/2}\|f\|_{L^2}.$$

Hence

$$\begin{aligned}\|f\|_{L^{2n/(n-2)}} &\leq \|a\|_{L^{2n/(n-2)}} + \|f-a\|_{L^{2n/(n-2)}} \\ &\leq |a|(\operatorname{Vol} M)^{(n-2)/(2n)} + \|f-a\|_{L^{2n/(n-2)}} \\ &\leq (\sqrt{2}+1)\,\|f-a\|_{L^{2n/(n-2)}} + \sqrt{2}\,(\operatorname{Vol} M)^{-1/n}\|f\|_{L^2}\end{aligned}$$

by (A.3.1). We have $\lambda_1^{1/2}\|f\|_{L^2} \leq \|df\|_{L^2}$ for $\int_M f = 0$, where λ_1 is the first positive eigenvalue of $\square$. We will show that λ_1 admits a lower bound of the form $\kappa_1(n,K_0)\,\frac{1}{D_g^2}$. Then from $\|dh\|_{L^2} \geq C_0\,(\frac{n-2}{n})\,\|h\|_{L^{2n/(n-2)}}$ we get our Sobolev inequality $\|df\|_{L^2} \geq C\|f\|_{L^{2n/(n-2)}}$ for $\int_M f = 0$ with $C = \left[(\sqrt{2}+1)\,C_0^{-1}(\frac{n}{n-2}) + \sqrt{2}\,(\operatorname{Vol} M)^{-1/n}\,\lambda_1^{-1/2}\right]^{-1}$. Now we try to prove the Sobolev inequality $\|df\|_{L^1} \geq C_0\|f\|_{L^{n/(n-1)}}$ when the volume of $\{f \leq 0\}$ equals the volume of $\{f \geq 0\}$.

(A.4) *Relation Between Sobolev Inequality and Isoperimetric Constant.* In the Sobolev inequality $\|df\|_{L^1} \geq C_0\|f\|_{L^{n/(n-1)}}$ when the volume of $\{f \leq 0\}$ equals the volume of $\{f \geq 0\}$, let us consider first the special case $f = \chi_{M_1} - a$, where χ_{M_1} is the characteristic function of an open subset

M_1 of M with smooth boundary S and a is a real number. Clearly a must be between 0 and 1. Let M_2 be the complement of M_1 in M. Then $\int_M |df| = \mathrm{Vol}(S)$ (when the integral is suitably interpreted in the sense of distributions) and

$$\left[\int_M |f|^{n/(n-1)}\right]^{(n-1)/n} = \left[(1-a)\mathrm{Vol}(M_1) + a\ \mathrm{Vol}(M_2)\right]^{(n-1)/n}.$$

If we have the Sobolev inequality, then

$$\mathrm{Vol}(S) \geq C_0\left[\mathrm{Min}(\mathrm{Vol}(M_1),\mathrm{Vol}(M_2))\right]^{(n-1)/n}.$$

This kind of inequality is known as an isoperimetric inequality. Conversely this isoperimetric inequality implies the Sobobev inequality. Let us formulate it more precisely. Let

$$\phi(M) = \inf_S \frac{(\mathrm{Vol}(S))^n}{(\mathrm{Min}(\mathrm{Vol}(M_1),\mathrm{Vol}(M_2)))^{n-1}},$$

where S runs over codimension one submanifolds of M which divide M into two pieces M_1 and M_2. To make referring to it easier we call $\phi(M)$ the *second isoperimetric constant* of M. We have seen that $\phi(M) \geq C_0^n$. We claim that conversely $\phi(M) \leq (2\ C_0)^n$.

Let f be a function on M and let $M_- = \{f \leq 0\}$ and $M_+ = \{f \geq 0\}$. We assume that $\mathrm{Vol}\ M_- = \mathrm{Vol}\ M_+$. We want to show that $\|f\|_{L^{n/(n-1)}(M)} \leq 2\ \phi(M)^{-1/n} \|df\|_{L^1(M)}$. For $t \geq 0$ let G_t be the set where $f \leq t$ and let M_t be the set where $f \geq t$ and S_t be the set where $f = t$. Since M_t is contained in $\{f \geq 0\}$ whose volume is equal to half of the volume of M, it follows that $\mathrm{Vol}(M_t) \leq \mathrm{Vol}(G_t)$ and

$\mathrm{Vol}(S_t) \geq \Phi(M)^{1/n}\,\mathrm{Vol}(M_t)^{(n-1)/n}$. Let $f_t = f\,\chi_{G_t}$ be the function obtained from f by truncation at the height t, where χ_{G_t} is the characteristic function of G_t. Let

$$u(t) = \left[\int_{G_t} |f|^{n/(n-1)}\right]^{(n-1)/n} = \|f_t\|_{L^{n/(n-1)}}.$$

Since $|f_{t+h}| \leq |f_t| + h\,\chi_{M_t}$ for $h > 0$, it follows that

$$\begin{aligned} u(t+h) &\leq u(t) + \|h\,\chi_{M_t}\|_{L^{n/(n-1)}} \\ &= u(t) + h\,\mathrm{Vol}(M_t)^{(n-1)/n} \\ &\leq u(t) + h\,\Phi(M)^{-1/n}\,\mathrm{Vol}(S_t) \end{aligned}$$

and $u'(t) \leq \Phi(M)^{-1/n}\,\mathrm{Vol}(S_t)$. Let t_* be the supremum of f on M. Since $u(0) = \|f\|_{L^{n/(n-1)}(M_-)}$, it follows that we have

$$\begin{aligned} \|f\|_{L^{n/(n-1)}(M)} &= u(t_*) = \|f\|_{L^{n/(n-1)}(M_-)} + \int_{t=0}^{t_*} u'(t)dt \\ &\leq \|f\|_{L^{n/(n-1)}(M_-)} + \Phi(M)^{-1/n} \int_{t=0}^{t_*} \mathrm{Vol}(S_t)\,dt \\ &= \|f\|_{L^{n/(n-1)}(M_-)} + \Phi(M)^{-1/n}\,\|df\|_{L^1(M_+)}. \end{aligned}$$

By replacing f by $-f$, we conclude that

$$\|f\|_{L^{n/(n-1)}(M)} \leq \|f\|_{L^{n/(n-1)}(M_+)} + \Phi(M)^{-1/n}\,\|df\|_{L^1(M_-)}.$$

Adding the two inequalities up, we get

$$\|f\|_{L^{n/(n-1)}(M)} \leq 2\,\Phi(M)^{-1/n}\,\|df\|_{L^1(M)}.$$

(A.5) *Lower Bound of First Eigenvalue of the Laplacian.* The lower bound of the first eigenvalue of the Laplacian is related to another isoperimetric constant which we call the *first isoperimetric constant* to make referring to it easier and which we denote by $I(M)$. It is defined as follows.

Suppose M is a compact manifold and is divided by a hypersurface S into two pieces M_1 and M_2. The first isoperimetric constant $I(M)$ is defined as the infimum of $\frac{\mathrm{Vol}\, S}{\min(\mathrm{Vol}\, M_1,\ \mathrm{Vol}\, M_2)}$ as S varies in the set of all hypersurfaces of M.

A theorem of Cheeger [Ch] says that the first eigenvalue λ_1 of the positive Laplacian $\square$ for functions is bounded from below by $\frac{1}{4}\, I(M)^2$. Let us now prove Cheeger's theorem. Let f be an eigenfunction for $\square$ for the first eigenvalue λ_1. Let M_+ be thet set where $f \geq 0$. By replacing f by $-f$ if necessary, we can assume without loss of generality that $\mathrm{Vol}(M_+) \leq \frac{1}{2}\mathrm{Vol}(M)$. For $t \geq 0$ let M_t be the set where $f^2 \geq t$ and S_t be the set where $f^2 = t$. Since M_t is contained in M_+ whose volume is no more than half of the volume of M, it follows that $\mathrm{Vol}(M_t) \leq \mathrm{Vol}(M - M_t)$ and $\mathrm{Vol}(S_t) \geq I(M)\,\mathrm{Vol}(M_t)$. Let t_* be the supremum of f on M. By using Lebesque's definition of an integral we have

$$\begin{aligned}\int_{M_+} f^2 &= -\int_{t=0}^{t_*} t\left[\frac{d}{dt}\mathrm{Vol}(M_t)\right]dt = -\,t\,\mathrm{Vol}(M_t)\Big|_{t=0}^{t_*} + \int_{t=0}^{t_*}\mathrm{Vol}(M_t)dt \\ &= \int_{t=0}^{t_*}\mathrm{Vol}(M_t)dt \leq \frac{1}{I(M)}\int_{t=0}^{t_*}\mathrm{Vol}(S_t)dt = \frac{1}{I(M)}\int_{M_+}|df^2| \\ &= 2\,\frac{1}{I(M)}\int_{M_+}|f|\;|df| \leq 2\,\frac{1}{I(M)}\left[\int_{M_+}|f|^2\right]^{1/2}\left[\int_{M_+}|df|^2\right]^{1/2}.\end{aligned}$$

Hence

$$\int_{M_+} f^2 \leq 4 \frac{1}{I(M)^2} \int_{M_+} |df|^2 = 4 \frac{1}{I(M)^2} \int_{M_+} \langle \Box f, f\rangle = 4 \frac{1}{I(M)^2} \lambda_1 \int_{M_+} |f|^2$$

and $\lambda_1 \geq \frac{1}{4} I(M)^2$.

(A.6) *Lower Bounds of Isoperimetric Constants.* We are left with the task of getting the lower bound estimates for the two isoperimetric constants. These lower bound estimates are given by a theorem of Gallot [G1,G2,B-B-G]. There exist positive constants $\beta_\upsilon(n,K_0)$, $\upsilon = 1, 2$, depending only on n and K_0 such that $I(M) \geq \beta_1(n,K_0) \frac{1}{D_g}$ and $\Phi(M) \geq \beta_2(n,K_0) \frac{V_g}{D_g^n}$. To give a detailed proof of these estimates would take us too far afield into geometric measure theory and differential geometry. So we give here only an indication of the proof. The proof for the lower bound of $\Phi(M)$ is analogous to that for $I(M)$. By rescaling we can assume without loss of generality that the diameter D_g is 1.

Step 1. For a positive number η there exists a domain M_1 in M such that the volume of M_1 is η and the volume of the boundary S of M_1 is the minimum among all boundaries of domains in M with volume equal to η. The set S_0 of regular points of S is of measure zero in S and the image of the exponential map defined by geodesics normal to S_0 contains $M - S$.

Step 2. By using the first variation of S_0 subject to the condition that the volume of M_1 is equal to η, one concludes that the mean curvature of S_0 is constant. We can assume that the mean curvature (calculated with respect to the outward normal of M_1) is nonnegative, otherwise we replace M_1 by its complement in M.

Step 3. We use the comparison theorem of Heintze and Karcher [H-K, p. 458, Corollary 3.3.2] and the constancy of the mean curvature of S_0 to estimate the volume of M_1 in terms of the volume of S_0. We conclude that the estimate is of the form $I(M) \geq \beta_1(n,K_0)\frac{1}{D_g}$ by using $D_g = 1$ and rescaling at the end of the process.

Remark. One can use also the lower bounds obtained by Croke [Cr] for $I(M)$ and $\Phi(M)$ using the integral geometry formula of Santalo [Sa, pp. 336-338] to get a lower bound for the Green's function. However, the lower bound so obtained is only of the form $-\gamma \frac{D_g^{2(n+1)}}{V_g^3}$, which is not good enough for our purpose.

CHAPTER 4. OBSTUCTIONS TO THE EXISTENCE OF KÄHLER-EINSTEIN METRICS

In this chapter we discuss the two obstructions to the existence of Kähler-Einstein metrics for the case of positive anticanonical class. The first one is the non-reductivity of the automorphism group discovered by Matsushima [M] and Lichnerowicz [Li]. the second one is the nonvanishing of an invariant for holomorphic vector fields due to Kazdan, Warner [K-W], and Futaki [F].

§1. *Reductivity of Automorphism Group.*

(1.1) First we discuss the Killing vector fields of a Riemannian manifold. A killing vector field is a vector field that leaves the Riemannian metric invariant. Let g_{ij} be the Riemannian metric and X^i be a vector field. Let $y^i = \varphi^i(x,t)$ be the 1-parameter subgroup obtained by integrating X^i. The Riemannian metric g_{ij} becomes

$$\begin{aligned} & g_{k\ell}(\varphi(x,t))\partial_i\varphi^k(x,t)\partial_j\varphi^\ell(x,t) \\ &= (g_{k\ell}(x) + t\partial_p g_{k\ell}(x)X^p)(\delta_i^k + t\partial_i X^k)(\delta_j^\ell + t\partial_j X^\ell) + o(t^2) \\ &= g_{ij} + t((\partial_p g_{ij})X^p + g_{kj}\partial_i X^k + g_{i\ell}\partial_j X^\ell) + o(t^2). \end{aligned}$$

Thus X^i is Killing if and only if the Lie derivative $(\partial_p g_{ij})X^p + g_{kj}\partial_i X^k + g_{i\ell}\partial_j X^\ell$ of g_{ij} vanishes. If η is the 1-form obtained by lowering the index of X, then the condition for X to be Killing becomes $\nabla_i\eta_j + \nabla_j\eta_i = 0$.

(1.2) Let us now assume that we have a compact complex manifold M whose anticanonical line bundle is positive and which carries a Kähler-Einstein metric. Recall that from the Bochner-Kodaira formula (1.5.1) of Chapter 3 every holomorphic vector field X^i is given by $\uparrow\bar\partial f$ for some eigenfunction f for the Laplacian $\Box$ with $\Box f = f$. We decompose f into its real part φ and its imaginary part ψ so that $f = \varphi + \sqrt{-1}\psi$. Let $Y^i = \uparrow\bar\partial\varphi$.

Consider $\operatorname{Im}Y^i$ which is given by $(-\frac{\sqrt{-1}}{2}Y^i, \frac{\sqrt{-1}}{2}Y^{\bar{i}})$.

We claim that $\operatorname{Im}Y^i$ is Killing. Its associated form η satisfies $\eta_i = \frac{\sqrt{-1}}{2}\partial_i\varphi$ and $\eta_{\bar{i}} = -\frac{\sqrt{-1}}{2}\partial_{\bar{i}}\varphi$. To verify that $\operatorname{Im}Y^i$ is Killing, we have to check that $\nabla_i\eta_j + \nabla_j\eta_i$, $\nabla_{\bar{i}}\eta_j + \nabla_j\eta_{\bar{i}}$, $\nabla_{\bar{i}}\eta_{\bar{j}} + \nabla_{\bar{j}}\eta_{\bar{i}}$ all vanish. Since φ is an eigenfunction of $\square$ associated to the eigenvalue 1 and since M is Kähler-Einstein, from the Bochner-Kodaira formula (1.5.1) of Chapter 3 applied to $\bar{\partial}\varphi$, it follows that $\bar{\nabla}\bar{\partial}\varphi = 0$. Because φ is real, we have also $\nabla\partial\varphi = 0$. Hence $\nabla_i\eta_j = \frac{\sqrt{-1}}{2}\nabla_i\partial_j\varphi = 0$ and $\nabla_{\bar{i}}\eta_{\bar{j}} = -\frac{\sqrt{-1}}{2}\nabla_{\bar{i}}\partial_{\bar{j}}\varphi = 0$. The remaining condition $\nabla_{\bar{i}}\eta_j + \nabla_j\eta_{\bar{i}} = 0$ follows from

$$\nabla_{\bar{i}}\eta_j + \nabla_j\eta_{\bar{i}} = \frac{\sqrt{-1}}{2}(\nabla_{\bar{i}}\partial_j\varphi - \nabla_{\bar{j}}\partial_i\varphi) = 0.$$

Now $\operatorname{Im}Y = -\operatorname{Re}(\sqrt{-1}Y) = -J\operatorname{Re}Y$ and $\operatorname{Re}Y = J\operatorname{Im}Y$. Let $Z = \uparrow\bar{\partial}\psi$. Then $\operatorname{Im}Z = -\operatorname{Re}(\sqrt{-1}Z)$ and $\operatorname{Re}X = \operatorname{Re}Y + \operatorname{Re}(\sqrt{-1}Z) = J\operatorname{Im}Y - \operatorname{Im}Z$. Since both $\operatorname{Im}Y$ and $\operatorname{Im}Z$ are Killing vector fields, it follows that every holomorphic vector field (which is identified with its real part when group action is considered) is a sum of a Killing vector field and the J of another Killing vector field. Also both Killing vector fields are the real parts of holomorphic vector fields (for the special case f is purely imaginary). Since the isometry group of a compact Riemannian manifold is a compact Lie group, it follows that the Lie algebra of the automorphism group of a compact Kähler-Einstein manifold with positive anticanonical line bundle is the complexification of the Lie algebra of a real compact Lie subgroup of the automorphism group. In other words, its automorphism group is *reductive*. We use this as our definition of reductivity. So we have the following theorem.

Theorem. The automorphism group of a compact Kähler-Einstein manifold with positive anticanonical line bundle is reductive.

An example of a compact Kähler manifold M with positive anticanonical line bundle whose automorphism group is not reductive is the complex projective space of complex dimension two with two points blown up. Let

$[z_0, z_1, z_2]$ be the homogeneous coordinates of $\mathbb{P}_2$ and let the two points to be blown up are on $z_0 = 0$. Then the translation group

$$[z_0, z_1, z_2] \to [z_0, z_1 + a_1 z_0, z_2 + a_2 z_0] \qquad (a_1, a_2 \in \mathbb{C})$$

defines a subgroup of the automorphism group of M which is biholomorphic to the additive group $\mathbb{C}^2$. The additive group $\mathbb{C}^2$ contains no compact subgroup.

§2. *The obstruction of Kazdan-Warner.*

(2.1) The Monge-Ampère equation for the existence of Kähler-Einstein metrics on a compact manifold of positive anticanonical class in the case of a complex curve is reduced to a quasi-linear equation which is of the same type as the equation to find a metric on the two-sphere which is conformal to the standard metric and whose Gaussian curvature is a prescribed function. Kazdan and Warner [K-W] discovered an obstruction to the solvability of the equation for a conformal metric with prescribed Gaussian curvature. This obstruction later led to the discovery by Futaki [F] of an invariant of holomorphic vector fields in the higher-dimensional case whose nonvanishing is an obstruction to the existence of a Kähler-Einstein metric. Later the invariant of Futaki was interpreted by Futaki and Morita [F-M] in terms of equivariant Chern class. We now first discuss the obstruction of Kazdan and Warner and its reformulation which relates it to the Futaki invariant.

First let us look at the equation for conformal metric with prescribed Gaussian curvature. On S^2 take a Hermitian metric $ds_0^2 = \sigma(z)dzd\bar{z}$. Its Gaussian curvature is $\frac{1}{2}K_0$, where $K_0 = -\frac{\Delta^0 \log\sigma}{\sigma}$ and $\Delta^0 = 4\frac{\partial^2}{\partial z \partial\bar{z}}$. The problem we want to consider is the following. Given a function K on S^2 we want to find a conformal factor λ so that $\frac{1}{2}K$ is the Gaussian curvature of the metric $ds^2 = \lambda ds_0^2$. Now

$$K = -\frac{\Delta^0 \log(\lambda\sigma)}{\lambda\sigma} = -\frac{\Delta^0 \log\sigma}{\lambda\sigma} - \frac{\Delta^0 \log\lambda}{\lambda\sigma} = \frac{1}{\lambda}(K_0 - \Delta\log\lambda)$$

where Δ is the Laplacian with respect to the metric ds_0^2. Hence

$$\lambda K = K_0 - \Delta\log\lambda.$$

Let $u = \log\lambda$. Then $\Delta u = K_0 - Ke^u$. For the standard metric ds_0^2 on S^2 the Gaussian curvature K_0 is identically 2. So in that case the equation for u becomes $\Delta u = 2 - Ke^u$.

(2.2) The key step to get the obstruction of Kazdan-Warner for the solvability of the equation $\Delta u = 2 - Ke^u$ is to consider the divergence of the function $2(\nabla f\cdot\nabla u)\nabla u - |\nabla u|^2\nabla f$ (where f is a smooth function on S^2). We have

$$\begin{aligned}\nabla\cdot(2(\nabla f\cdot\nabla u)\nabla u - |\nabla u|^2\nabla f) &= 2H_f(\nabla u,\nabla u) + 2H_u(\nabla f,\nabla u) + 2\Delta u(\nabla f\cdot\nabla u)\\ &\quad - 2H_u(\nabla u,\nabla f) - |\nabla u|^2\Delta f,\end{aligned}$$

where H_f and H_u mean respectively the Hessians of f and u. Hence

$$2\Delta u(\nabla f\cdot\nabla u) = \nabla\cdot(2(\nabla f\cdot\nabla u)\nabla u - |\nabla u|^2\nabla f) - (2H_f - (\Delta f)ds_0^2)(\nabla u,\nabla u).$$

The obstruction comes from eigenfuctions of the Laplacian for the eigenvalue -1. As we saw earlier, these eigenfunctions are related to holomorphic vector fields in the case of Kähler-Einstein metrics. The eigenfunctions of the Laplacian for the sphere can be described by spherical harmonics. So let us now look at the spherical harmonics. Let us do this in the case of the n-sphere.

For a sphere S^n in $\mathbb{R}^{n+1}$ we have the following relation between the Laplacians of S^n and $\mathbb{R}^{n+1}$ in polar coordinates by using the divergence theorem.

$$\Delta_{\mathbb{R}^{n+1}} = \frac{\partial^2}{\partial r^2} + \frac{n}{r}\frac{\partial}{\partial r} + \frac{1}{r^2}\Delta_{S^n}.$$

More precisely, take a region D in S^n and take the cone with base D and vertex 0. Cut the cone by the spheres $r = r_0$ and $r = r + \delta r$. Apply the divergence theorem on the part of the cone between the two spheres. Then the divergence of v as D shrinks to a point and $\delta r \to 0$ is the limit of

$$\frac{1}{(\mathrm{Vol}(D))r^n\delta r}\left[(r^n(\mathrm{Vol}(D))\frac{\partial v}{\partial r})\Big|_{r=r_0}^{r=r_0+\delta r} + \delta r\int_{\partial D_r}\frac{\partial v}{\partial \vec{n}}\right],$$

where D_r is the intersection of the cone with the sphere of radius r and $\vec{n}$ is the outward normal of ∂D_r along the sphere S^n_r of radius r. Hence the divergence of v is equal to

$$\frac{1}{r^n}\frac{\partial}{\partial r}(r^n\frac{\partial v}{\partial r}) + \Delta_{S^n_r}v = \frac{\partial^2 v}{\partial r^2} + \frac{n}{r}\frac{\partial v}{\partial r} + \frac{1}{r^2}\Delta_{S^n}v.$$

Thus any harmonic polynormial of degree d when restricted to the unit sphere S^n is an eigenfunction of Δ_{S^n} with eigenvalue $-d(n+d-1)$. In particular, for a linear spherical harmonic f in the case of $n = 2$ we have $\Delta f = -2f$. Moreover, $2H_f - (\Delta f)ds_0^2 = 0$ (that is, the Hessian of f is diagonal), because the restriction of f to any great circle of S^2 is an eigenfunction of the Laplacian of the great circle with eigenvalue -1.

(2.3) We now use the notation ≈ 0 to mean modulo divergence terms. Then for the restriction f of any linear spherical harmonic to S^2 and for any smooth function u on S^2, we have $\Delta u(\nabla f\cdot\nabla u) \approx 0$. In particular, if u is a solution of $\Delta u = 2 - Ke^u$, we have $(2 - Ke^u)(\nabla f\cdot\nabla u) \approx 0$ and

$$\begin{aligned}2\nabla f\cdot\nabla u &\approx Ke^u(\nabla f\cdot\nabla u) \approx K(\nabla e^u)\cdot\nabla f \approx -e^u\nabla\cdot(K\nabla f)\\ &\approx - e^u\nabla K\cdot\nabla f - e^uK\Delta f \approx - e^u\nabla K\cdot\nabla f + 2e^uKf.\end{aligned}$$

On the other hand,

$$2\nabla f\cdot\nabla u \approx -2(\Delta u)f = -2(2 - Ke^{u})f \approx 2Ke^{u}f,$$

because f is itself a divergence term. Hence we have the final conclusion that *if* u *is a solution of the equation* $\Delta u = 2 - Ke^{u}$ *and* f *is a linear spherical harmonic, then*

$$\int_{S^2} e^{u}\nabla K\cdot\nabla f = 0.$$

In particular, when $K = K_0 + f$ we have $\int_{S^2} K = \int_{S^2} K_0$, but

$$\int_{S^2} e^{u}\nabla K\cdot\nabla f = \int_{S^2} e^{u}|\nabla f|^2 > 0.$$

(2.4) Now we reinterprete this obstruction of Kazdan-Warner to motivate the definition of the invariant of Futaki. The equation $\Delta u = 2 - Ke^{u}$ can be rewritten as $1 + (\frac{1}{2}\Delta)(-u) = \frac{K}{2}e^{-(-u)}$. So when we replace u by $-u$ and use the complex Laplacian (which is equal to one-half of the real Laplacian), we get

$$1 + \Delta u = e^{-u+F},$$

where $F = \log\frac{K}{2}$. (Note that though $\Delta = 4\dfrac{\partial^2}{\partial z\partial\bar{z}}$ we have only a factor of two when we go to the complex Laplacian because there is a factor of two in the expression relating the real and the complex metrics.) The function F now is associated to the eigenvalue -1 instead of -2, because of the change of the meaning of the Laplacian from the real to the complex. The Kazdan-Warner obstruction becomes

$$\int_{S^2} e^{-u+F}\nabla F\cdot\nabla f = 0.$$

For a Kähler-Einstein metric the holomorphic vector fields X are given by $\uparrow\bar{\partial}F$. So we have

$$\int_{S^2} e^{-u+F} XF = 0$$

for any holomorphic vector field X on S^2.

Even though we are only in the case of complex dimension one, we use the language of indices to see what the higher dimensional analog should be. We let $g'_{i\bar{j}} = g_{i\bar{j}} + \partial_i\partial_{\bar{j}}u$, where $g_{i\bar{j}}$ is the standard metric on S^2. Then the volume form of $g'_{i\bar{j}}$ is e^{-u+F} times the volume form of $g_{i\bar{j}}$. Also we have

$$R'_{i\bar{j}} = R_{i\bar{j}} + \partial_i\partial_{\bar{j}}u - \partial_i\partial_{\bar{j}}F = R_{i\bar{j}} + g'_{i\bar{j}} - g_{i\bar{j}} - \partial_i\partial_{\bar{j}}F$$

and

$$R'_{i\bar{j}} - g'_{i\bar{j}} = -\partial_i\partial_{\bar{j}}F.$$

So we have the vanishing of the integral of XF over S^2 with respect to the volume form of $g'_{i\bar{j}}$. This integral will be precisely the Futaki invariant we are about to define.

§3. *The Futaki Invariant*

Let M be a compact Kähler manifold with positive anticanonical line bundle K_M^{-1}. Choose a metric $g_{i\bar{j}}$ in the anticanonical class. Let $R_{i\bar{j}}$ be its Ricci curvature tensor. There exists a real-valued function f unique up to an additive constant such that $R_{i\bar{j}} - g_{i\bar{j}} = \partial_i\partial_{\bar{j}}F$. Let X be a holomorphic vector field on M. The invariant of Futaki for the holomorphic vector field X is $\int_M (XF)dV(g)$. where $dV(g)$ is the volume for the Kähler metric $g_{i\bar{j}}$. The important property of the invariant of Futaki is that it is independent of the Kähler metric $g_{i\bar{j}}$ chosen. We are going to look at this independence from the point of view of eigenfunctions.

We introduce the metric e^F on the trivial line bundle over M and also multiply the usual metric of the anticanonical line bundle by the factor e^F. Then the new curvature form of the anticanonical line bundle is $R^F_{i\bar{j}} = R_{i\bar{j}} - \partial_i\partial_{\bar{j}}F$ which is equal to the Kähler metric $g_{i\bar{j}}$. We are going to use the subscript or superscript F to denote tensors or operators associated with this new metric obtained by twisting the usual metric by the factor e^F.

Take a function η on M and consider the following Bochner-Kodaira formula for the (0,1)-form $\bar{\partial}\eta$.

$$\square_F\bar{\partial}\eta = -g^{i\bar{j}}\nabla^F_i\nabla_{\bar{j}}\bar{\partial}\eta + R^F\cdot\bar{\partial}\eta,$$

where $\square_F = \bar{\partial}^*_F\bar{\partial} + \bar{\partial}\bar{\partial}^*_F$. Since the operator R^F is simply the identity, we have $(\square_F - 1)\bar{\partial}\eta = -g^{i\bar{j}}\nabla^F_i\nabla_{\bar{j}}\bar{\partial}\eta$. If $\nabla_{\bar{j}}\bar{\partial}\eta = 0$ (i.e. $\uparrow\bar{\partial}\eta$ is a holomorphic vector field), then $(\square_F - 1)\bar{\partial}\eta = 0$ which implies that $\bar{\partial}(\square_F - 1)\eta = 0$ and $(\square_F - 1)\eta = \text{constant}$. Given any holomorphic vector field X on M, the (0,1)-form $g_{i\bar{j}}X^i$ is $\bar{\partial}$-closed. Since the anticanonical line bundle is positive, by Kodaira's vanishing theorem the cohomology group $H^1(M,\mathcal{O}_M) = 0$ and every $\bar{\partial}$-closed (0,1)-form is $\bar{\partial}$-exact. So every holomorphic vector field X is of the form $\uparrow\bar{\partial}\eta$ for some smooth function η on M. We have seen that $(\square_F - 1)\eta = \text{constant}$ and by modifying η by an additive constant, we can assume that $(\square_F - 1)\eta = 0$. Conversely, if $(\square_F - 1)\eta = 0$, then $g^{i\bar{j}}\nabla^F_i\nabla_{\bar{j}}\bar{\partial}\eta = 0$ and

$$\int_M |\bar{\nabla}\bar{\partial}\eta|^2 e^{-F}dV(g) = \int_M -\langle g^{i\bar{j}}\nabla^F_i\nabla_{\bar{j}}\bar{\partial}\eta, \bar{\partial}\eta\rangle e^{-F}dV(g) = 0$$

and $\bar{\nabla}\bar{\partial}\eta = 0$ and $\uparrow\bar{\partial}\eta$ is a holomorphic vector field. Thus we have the following statement essentially due to Matsushima. $\uparrow\bar{\partial}\eta$ is a holomorphic vector field if and only if $(\square_F - 1)\eta = 0$ after modifying η by an additive constant.

Let $X = \uparrow\bar{\partial}\eta$ be a holomorphic vector field on M. Then for any function ψ on M, $X\psi = g^{i\bar{j}}(\partial_{\bar{j}}\eta)\partial_i\psi = \langle\bar{\partial}\eta,\bar{\partial}\psi\rangle$. We also note that $\square_F\eta = -\Delta\eta + \langle\bar{\partial}\eta,\bar{\partial}F\rangle$.

Suppose now we have a smooth family of metrics $g_{i\bar{j}}$ depending on a real parameter t. We use an overhead dot $\dot{}$ to denote differentiation with respect to t. By differentiating $R_{i\bar{j}} - g_{i\bar{j}} = \partial_i\partial_{\bar{j}}F$ with respect to t, we get $\partial_i\partial_{\bar{j}}(-\Delta\dot{\varphi}) - \partial_i\partial_{\bar{j}}\dot{\varphi} = \partial_i\partial_{\bar{j}}\dot{F}$. Since for every fixed t, F is determined up to a constant, we can choose the constant so that $-\Delta\dot{\varphi} - \dot{\varphi} = \dot{F}$. Thus

$$\frac{d}{dt}\int_M (XF)dV(g) = \int_M \left[X(-\Delta\dot{\varphi} - \dot{\varphi}) + (XF)\Delta\dot{\varphi}\right]dV(g),$$

because the derivative of $dV(g)$ with respect to t is $(\Delta\dot{\varphi})dV(g)$. It follows that

$$\begin{aligned}\frac{d}{dt}\int_M (XF)dV(g) &= \int_M \left[\langle\bar{\partial}\eta,\bar{\partial}(-\Delta\dot{\varphi} - \dot{\varphi})\rangle + \langle\bar{\partial}\eta,\bar{\partial}F\rangle\Delta\dot{\varphi}\right]dV(g)\\ &= \int_M \left[\langle-\Delta\eta,(-\Delta\dot{\varphi} - \dot{\varphi})\rangle + \langle\bar{\partial}\eta,\bar{\partial}F\rangle\Delta\dot{\varphi}\right]dV(g)\\ &= \int_M \left[\langle-\Delta\eta - \eta,-\Delta\dot{\varphi}\rangle + \langle\bar{\partial}\eta,\bar{\partial}f\rangle\Delta\dot{\varphi}\right]dV(g)\\ &= \int_M \langle\square_F\eta - \eta,-\Delta\dot{\varphi}\rangle dV(g)\end{aligned}$$

which is zero, because $\square_F\eta - \eta \equiv 0$ on M. The computation in the verification given here of the independence of Futaki's invariant on the choice of the Kähler metric is less involved than in Futaki's original proof and is due to Pit-Mann Wong.

An example of a compact Kähler manifold with positive anticanonical line bundle whose automorphism group is reductive and yet whose Futaki invariants are not all zero is the following. Take the rank two bundle E over $\mathbb{P}_1 \times \mathbb{P}_2$ which is the direct sum of the hyperplane bundle over $\mathbb{P}_1$ and the hyperplane bundle over $\mathbb{P}_2$. The total space of the projective bundle of E is an example. This example is given in [F, p.440, Prop.3] and is verified there. The verification is computationally rather involved. We are not reproducing it here.

CHAPTER 5. MANIFOLDS WITH SUITABLE FINITE SYMMETRY

§1. *Motivation for the use of finite symmetry.*

In Chapter 4 we discussed two obstructions to the existence of Kähler-Einstein metrics on compact Kähler manifolds with positive anticanonical line bundle. These two obstructions are related to the presence of nonzero holomorphic vector fields. There is a conjecture that any compact Kähler manifold of positive anticanonical line bundle without any nonzero holomorphic vector field admits a Kähler-Einstein metric. The conjecture is still open. The only known examples of Kähler-Einstein metrics of Kähler manifolds of positive anticanonical line bundle are those of Hermitian symmetric manifolds or homogeneous manifolds or certain noncompact manifolds [C5]. So far there is no known way of proving the existence of Kähler-Einstein metrics of compact Kähler manifolds of positive anticanonical line bundle by using the continuity method with reasonable additional assumptions such as the nonexistence of nonzero holomorphic vector fields. In this Chapter we discuss a method [Siu1,Siu2] to prove the existence of Kähler-Einstein metrics for compact Kähler manifolds of positive anticanonical line bundle under the additional assumption of the existence of a suitable finite or compact group of symmetry. The method is not very satisfactory, because its applicability is exceedingly limited. This method can be applied to prove the existence of a Kähler-Einstein metric on the Fermat cubic surface and the surface obtained by blowing up three points of the complex projective surface $\mathbb{P}_2$. The method is also applicable to higher-dimensional Fermat hypersurfaces. We sketch in this chapter only the main ideas and the key steps of this method. Details can be found in [Siu2].

In the case of positive anticanonical line bundle the only difficulty in getting a Kähler-Einstein metric is the lack of a zeroth order *a priori* estimate for the solution of the Monge-Ampère equation by the continuity method. As discussed in §2 of Chapter 4 the one-dimensional case of the Monge-Ampère equation for the case of positive anticanonical line bundle is of the same type as the equation to find a metric on the two-sphere which is conformal to the standard metric and whose Gaussian curvature is a prescribed function. Moser [Mo3] proved that when there is antipodal symmetry for the prescribed function, the equation to to find a metric on the two-sphere which is conformal to the standard metric and whose Gaussian curvature is a prescribed function can be solved. This motivates the use of a finite group of symmetry to solve the Monge-Ampère equation for the case of positive anticanonical class. Properties of the finite group of symmetry will be used to get a zeroth order *a priori* estimate of the solution of the Monge-Ampère equation. The technical key step is to apply the simple inequality $uv \leq u\log u + e^{v-1}$ to the Green's formula for the restriction, to a complex curve, of the solution of the Monge-Ampère equation so that one can transform the product of the Green's kernel and the Laplacian on the curve of the solution of the Monge-Ampère equation into a sum. Since the Laplacian of the solution of the Monge-Ampère equation is bounded by the exponential of a constant times the difference of its supremum and infimum, one would get a zeroth order *a priori* estimate if the curve passes through a supremum point and an infimum point and the area of the curve is small relative to the constant in the exponent of the estimate of the Laplacian of the solution. The use of symmetry has the same effect as reducing the area of the curve by taking the quotient with respect to the finite group of symmetry. The constant in the exponent of the estimate of the Laplacian of the solution of the Monge-Ampère equation is linked to the lower bound of the bisectional curvature of the manifold. As a consequence the conditions on the finite group of symmetry is related to the lower bound of the bisectional curvature of the manifold for two orthonormal directions, the area of a (possibly reducible) curve joining two arbitrary points, and the number of points in a branch of the curve which are congruent under the group. Because in general the computation of a good explicit lower bound of the bisectional curvature for two orthonormal directions is rather difficult, in our applications we have to modify the argument so that the more easily computable bisectional curvature with a conformal factor is used instead of the usual bisectional curvature.

§2. *Relation Between* $\sup_M \varphi$ *and* $\inf_M \varphi$.

(2.1) Let M be a compact Kähler manifold of complex dimension m. Let $g_{i\bar{j}}$ be a Kähler metric of M whose Kähler form is in the anticanonical class of M. There exists a real-valued smooth function F on M with $\int_M e^F = \mathrm{Vol}\, M$ such that the Ricci curvature $R_{i\bar{j}}$ of $g_{i\bar{j}}$ satisfies

$$R_{i\bar{j}} - g_{i\bar{j}} = \partial_i \partial_{\bar{j}} F.$$

As in Chapter 3 we try to solve the Monge-Ampère equation

(2.1.1) $$\frac{\det(g_{i\bar{j}} + \partial_i \partial_{\bar{j}} \varphi)}{\det(g_{i\bar{j}})} = \exp(-t\varphi + F)$$

for the function $\varphi = \varphi_t$ on M $(0 \le t \le 1)$ by the continuity method. Again since the Ricci curvature $R'_{i\bar{j}}$ of the Kähler metric $g'_{i\bar{j}} = g_{i\bar{j}} + \partial_i \partial_{\bar{j}} \varphi$ satisfies

$$R'_{i\bar{j}} = t g'_{i\bar{j}} + (1-t) g_{i\bar{j}},$$

the metric $g'_{i\bar{j}}$ for $t = 1$ is a Kähler-Einstein metric.

As in Chapter 3 we have openness if for openness at $t = 0$ we consider instead the Monge-Ampère equation

$$\frac{\det(g_{i\bar{j}} + \partial_i \partial_{\bar{j}} \varphi)}{\det(g_{i\bar{j}})} = \left[\frac{1}{\mathrm{Vol}\, M} \int_M e^{-t\varphi + F}\right]^{-1} e^{-t\varphi + F}$$

with $\int_M \varphi = 0$. One gets in the same way as in Chapter 2 the first, second, and $(2+\epsilon)$ order *a priori* estimates for the function φ, provided that one has the zeroth order *a priori* estimate for φ. So the difficult part is the zeroth order *a priori* estimate for φ. Because of two known obstructions discussed in Chapter 4 we know that in general one cannot have the zeroth order *a priori* estimate for φ in the case of positive anticanonical line bundle. However, it is possible to estimate the supremum of $-\varphi$ (respectively φ) from above in terms of the supremum of φ (respectively $-\varphi$). The first of these two estimates is needed for our method. The precise statement of these two estimates are as follows.

(2.2) *Proposition.* Given any positive number ϵ and any $0 < t_0 \le 1$ there exists a positive constant C such that if φ is a solution of (2.1.1) on M for $0 \le t < t_0$, then $\sup_M(-\varphi) \le (m+\epsilon)\sup_M\varphi + C$ and $\sup_M\varphi \le (m+\epsilon)\sup_M(-\varphi) + C$ on M for $0 \le t < t_0$.

We sketch the proof of the first inequality in the conclusion of Proposition (2.2). The proof of the second inequality is analogous. By using $R'_{i\bar{j}} = tg'_{i\bar{j}} + (1-t)g_{i\bar{j}}$ and a Bochner type formula to get lower eigenvalue estimates, we obtain for $t+s > 0$ the Poincaré type inequality

$$\int_M |f|^2 e^{-s\varphi} dV' \le \frac{1}{t+s}\int_M \langle\bar{\partial}f,\bar{\partial}f\rangle' e^{-s\varphi} dV' + \frac{1}{t+s}\,\frac{\left[\int_M fe^{-s\varphi}dV'\right]^2}{\int_M e^{-s\varphi}dV'}$$

for any smooth function f on M, where dV' is the volume form of $g'_{i\bar{j}}$ and $\langle\cdot,\cdot\rangle'$ is the inner product with respect to $g'_{i\bar{j}}$. Using $\Delta'\varphi \le m$ and $f = e^{s\varphi}$ and Hölder's inequality, we get

$$\int_M e^{s\varphi - t\varphi + F} dV \le \frac{1}{1 + \frac{ms}{t+s}}\int_M e^{(t+s)\varphi - F} dV.$$

Choosing $s = -(\frac{1}{m+1} - \epsilon)t$ for some small positive number ϵ and using $\Delta'\varphi \le m$ and the fact that the Green's function for Δ' is bounded from below by an *a priori* constant, we obtain $\sup_M(-\varphi) \le (m+\epsilon)\ \sup_M\varphi + C$.

§3. *Estimate of* $m+\Delta\varphi$.

We need a second-order *a priori* estimate of φ which is a slightly modified form of that given in §3 of Chapter 2, because we would like to use the more easily computable bisectional curvature *with a conformal factor* instead of the usual bisectional curvature. The method used in §3 of Chapter 2 to establish the second-order *a priori* estimate of φ can easily be modified to give the form we want.

(3.1) *Definition.* Let $g_{i\bar{j}}$ be a Kähler metric of a complex manifold M and $R_{i\bar{j}k\bar{\ell}} = -\partial_i\partial_{\bar{j}}g_{k\bar{\ell}} + g^{\lambda\bar{\mu}}\partial_i g_{k\bar{\mu}}\partial_{\bar{j}}g_{\lambda\bar{\ell}}$ be its curvature tensor. Let ψ be a smooth real-valued function on M. We say that *with the conformal factor* $e^{-\psi}$ *the bisectional curvature of* $g_{i\bar{j}}$ *for two orthonormal vectors is bounded from below by* A if $(\psi_{i\bar{j}} + R_{i\bar{j}k\bar{\ell}})\xi^i\overline{\xi^j}\eta^k\overline{\eta^\ell} \geq A$ for any (ξ^i) and (η^i) that satisfy $g_{i\bar{j}}\xi^i\overline{\xi^j} = g_{i\bar{j}}\eta^i\overline{\eta^j} = 1$ and $g_{i\bar{j}}\xi^i\overline{\eta^j} = 0$.

(3.2) *Proposition.* Suppose $h_{i\bar{j}}$ is a Kähler metric in the anticanonical class of M and ψ is a smooth real-valued function on M. Assume that with the conformal factor $e^{-\psi}$ the bisectional curvature of $h_{i\bar{j}}$ for two orthonormal vectors is bounded from below by some real number $-\kappa$. Let γ be a nonnegative number $> \kappa$. Then there exist positive constants C and C' such that if φ is a solution of the Monge-Ampère equation (2.1.1) on M for $0 \leq t < t_0$, then $m+\Delta\varphi \leq C\exp(\gamma\varphi - (\gamma+1)\inf_M\varphi) + C'$ on M for $0 \leq t < t_0$. Here Δ means the (negative) Laplacian with respect to the Kähler metric $g_{i\bar{j}}$.

In our application the following conformal factor $e^{-\psi}$ is used. Suppose the bisectional curvature of $h_{i\bar{j}}$ for two orthonormal vectors is bounded from above by some real number κ_1 and the holomorphic sectional curvature of $h_{i\bar{j}}$ for unit vectors is bounded from above by some real number

κ_2. Suppose μ is a Hermitian metric along the fibers of the anticanonical line bundle K_M^{-1} whose curvature form dominates σ times $h_{i\bar{j}}$ for some real number σ. Choose a real-valued smooth function ψ on M so that $\partial_i\partial_{\bar{j}}\psi + R^h_{i\bar{j}}$ equals the curvature form of the Hermitian metric μ of K_M^{-1}, where $R^h_{i\bar{j}}$ is the Ricci curvature form of $h_{i\bar{j}}$. Then with the conformal factor $e^{-\psi}$ the bisectional curvature of $h_{i\bar{j}}$ for two orthonormal vectors is bounded from below by $\sigma - \kappa_2 - (m-2)\kappa_1$.

§4. *The use of a finite group of symmetry.*

(4.1) First let us make some simple remarks. On the complex line $\mathbb{C}$ with coordinate z our Laplacian $\Delta = g^{i\bar{j}}\partial_i\partial_{\bar{j}}$ becomes $\frac{\partial^2}{\partial z\partial\bar{z}}$. For any smooth function f on $\mathbb{C}$ with compact support the Cauchy integral formula for smooth functions gives

$$f(0) = \frac{1}{2\pi\sqrt{-1}}\int_{\mathbb{C}}\frac{\frac{\partial f}{\partial\bar{z}}(z)dz\wedge d\bar{z}}{z} = \frac{1}{2\pi\sqrt{-1}}\int_{\mathbb{C}}(\frac{\partial}{\partial z}\log|z|^2)\frac{\partial f}{\partial\bar{z}}(z)dz\wedge d\bar{z}$$

$$= \int_{\mathbb{C}}(\frac{1}{2\pi}\log|z|^2)\sqrt{-1}\partial\bar{\partial}f(z).$$

So the Green's function is $\frac{-1}{2\pi}\log|z|^2$. For the general case of a nonsingular complex curve Γ with a Kähler metric, the dominant term of the Green's function $G_\Gamma(x,y)$ for Γ is $\frac{-1}{2\pi}\log \operatorname{dist}(x,y)^2$ near $x = y$; and for any smooth function f on Γ we have

$$(4.1.1) \qquad f(x) = \frac{1}{\operatorname{Vol}(\Gamma)}\int_\Gamma f + \int_{y\in\Gamma} G_\Gamma(x,y)(-\sqrt{-1}\partial\bar{\partial}f)(y).$$

Consider now our compact Kähler manifold M of complex dimension m with a Kähler metric $g_{i\bar{j}}$ on which we would like to solve by the continuity method the Monge-Ampère equation (2.1.1). Assume that G is a finite subgroup of automorphisms of M. We require that the Kähler metric $g_{i\bar{j}}$ and the function F both be invariant under G. We consider only solutions φ that are invariant under G. We are going to discuss how the finite group of symmetry G would help us to get our zeroth-order *a priori* estimate of φ. To illustrate the idea, let us consider only the simplest situation. Take a nonsingular complex curve Γ in M which contains both a point P where $\sup_M \varphi$ is achieved and a point Q where $\inf_M \varphi$ is achieved. Let $\omega = \sqrt{-1} g_{i\bar{j}} dz^i \wedge \overline{dz^j}$ and $\omega' = \sqrt{-1} g'_{i\bar{j}} dz^i \wedge \overline{dz^j}$ be the Kähler forms of the two Kähler metrics $g_{i\bar{j}}$ and $g'_{i\bar{j}} = g_{i\bar{j}} + \partial_i \partial_{\bar{j}} \varphi$ on M. Let $G_\Gamma(x,y)$ be the Green's function for the Laplacian of the restriction of the Kähler metric $g_{i\bar{j}}$ to Γ. Then for any smooth function f on Γ, we have the Green's formula (4.1.1). Let K be a positive number such that $G_\Gamma(x,y) + K \geq 0$. Then

$$f(x) = \frac{1}{\mathrm{Vol}(\Gamma)} \int_\Gamma f\,\omega + \int_{y \in \Gamma} (G_\Gamma(x,y) + K)(-\sqrt{-1}\partial\bar{\partial} f)(y).$$

Let P_μ and Q_μ $(1 \leq \mu \leq N)$ be points of Γ so that Q_μ $(1 \leq \mu \leq N)$ are *distinct*. Assume that all the P_μ's are congruent to P under G and all the Q_μ's are congruent to Q under G. Since $\omega' = \omega + \sqrt{-1}\partial\bar{\partial}\varphi$ is positive definite, it follows that $-\sqrt{-1}\partial\bar{\partial}\varphi < \omega$ and $\sqrt{-1}\partial\bar{\partial}\varphi < \omega'$. Applying the above formula to φ and $-\varphi$ at the points P_μ and Q_μ respectively, we get

$$\begin{aligned}
\varphi(P_\mu) &= \frac{1}{\mathrm{Vol}(\Gamma)} \int_\Gamma \varphi\,\omega + \int_{y \in \Gamma} (G_\Gamma(P_\mu, y) + K)(-\sqrt{-1}\partial\bar{\partial}\varphi)(y). \\
&\leq \frac{1}{\mathrm{Vol}(\Gamma)} \int_\Gamma \varphi\,\omega + \int_{y \in \Gamma} (G_\Gamma(P_\mu, y) + K)\omega \\
&= \frac{1}{\mathrm{Vol}(\Gamma)} \int_\Gamma \varphi\,\omega + K\mathrm{Vol}(\Gamma).
\end{aligned}$$

$$\varphi(Q_\mu) = \frac{-1}{\mathrm{Vol}(\Gamma)}\int_\Gamma \varphi\,\omega + \int_{y\in\Gamma}(G_\Gamma(Q_\mu,y) + K)(\sqrt{-1}\partial\bar\partial\varphi)(y).$$
$$\leq \frac{-1}{\mathrm{Vol}(\Gamma)}\int_\Gamma \varphi\,\omega + \int_{y\in\Gamma}(G_\Gamma(Q_\mu,y) + K)\omega'(y).$$

We use the inequality $uv \leq u\log u + e^{v-1}$ with u equal to the quotient of $\frac{1}{2\pi-\epsilon}\omega'|\Gamma$ by $\omega|\Gamma$ and v equal to $(2\pi-\epsilon)(\Sigma_{\mu=1}^N G_\Gamma(Q_\mu,y) + NK)$. By using also the fact that the quotient of $\omega'|\Gamma$ by $\omega|\Gamma$ is $\leq m + \Delta\varphi$, we get

$$\Sigma_{\mu=1}^N -\varphi(Q_\mu) \leq \Sigma_{\mu=1}^N\frac{-1}{\mathrm{Vol}(\Gamma)}\int_\Gamma \varphi\,\omega + \int_{y\in\Gamma}(\Sigma_{\mu=1}^N G_\Gamma(Q_\mu,y) + NK)\omega'(y)$$
$$\leq \Sigma_{\mu=1}^N\frac{-1}{\mathrm{Vol}(\Gamma)}\int_\Gamma \varphi\,\omega + \frac{1}{2\pi-\epsilon}\left[\sup_\Gamma\log(m + \Delta\varphi) + \log\frac{1}{2\pi-\epsilon}\right]\mathrm{Vol}(\Gamma)$$
$$+ \int_{y\in\Gamma}\exp\left[(2\pi-\epsilon)(\Sigma_{\mu=1}^N G_\Gamma(Q_\mu,y) + NK)\right]\omega$$

Here ϵ is any small positive number. By adding together the inequalities for $\Sigma_{\mu=1}^N\varphi(P_\mu)$ and $\Sigma_{\mu=1}^N -\varphi(Q_\mu)$, we get

$$\Sigma_{\mu=1}^N\left[\varphi(P_\mu)-\varphi(Q_\mu)\right] \leq \frac{1}{2\pi-\epsilon}\left[\sup_\Gamma\log(m + \Delta\varphi)(y) + (2\pi-\epsilon)NK + \log\frac{1}{2\pi-\epsilon}\right]\mathrm{Vol}(\Gamma)$$
$$+ \int_{y\in\Gamma}\exp\left[(2\pi-\epsilon)(\Sigma_{\mu=1}^N G_\Gamma(Q_\mu,y) + NK)\right]\omega.$$

Finally we use $m+\Delta\varphi \leq C\exp(\gamma\varphi - (\gamma+1)\inf_M\varphi) + C'$ and $\varphi(P_\mu) = \sup_M\varphi$ and $\varphi(Q_\mu) = \inf_M\varphi$ to conclude that

$$N(\sup_M\varphi - \inf_M\varphi) \leq \frac{1}{2\pi-\epsilon}\left[\gamma\sup_M\varphi - (\gamma+1)\inf_M\varphi\right]\mathrm{Vol}(\Gamma) + C^\#$$

for some constant $C^\#$ independent of φ. So for example when N is greater than $\frac{\gamma+1}{2\pi}\mathrm{Vol}(\Gamma)$ we have a zeroth-order *a priori* estimate of φ. In actual applications the situation is more complicated and a sequence of possibly singular irreducible curves is required to join a supremum point of φ to an infimum point of φ in order to assure that there are enough congruent points on each irreducible curve with small volume to make the argument work.

The relation between the supremum of φ and the infimum of φ has to be used in the situation when no supremum point of φ and no infimum point of φ lie on the same irreducible curve with enough congruent points and small enough volume. To precisely state our main result we need some definitions.

(4.2) *Definition.* Let S be a connected complex manifold. A holomorphic family $\mathcal{F}$ of (possibly singular) complex curves Γ_s with base point O_s $(s \in S)$ is said to be *smoothly simultaneously uniformizable* if there exist
(i) a differentiable manifold $\mathcal{G}$ and a smooth submersion $\theta: \mathcal{G} \to S$ whose fibers are compact of real dimension 2 and
(ii) two smooth maps $\sigma: S \to \mathcal{G}$ and $\tau: \mathcal{G} \to M$ such that for every $s \in S$
(a) $\theta^{-1}(s)$ can be given a complex structure so that $\theta^{-1}(s)$ is the normalization of Γ_s under the map τ and
(b) $\tau(\sigma(s)) = O_s$.

(4.3) *Definition.* By the *curve volume* of the family $\mathcal{F}$ of holomorphic curves we mean the volume of the curve Γ_s with respect to the anticanonical class of M (which is independent of the choice of s in S and independent of the choice of the Kähler metric in the anticanonical class).

(4.4) *Definition.* We say that the *orbit cardinality* of the family $\mathcal{F}$ of holomorphic curves is at least N if there exists some positive number δ such that for every $s \in S$ there are at least N points $Q_1, \cdots, Q_N$ in the orbit GO_s (under the group G) of the base point O_s of the curve Γ_s with the distance between Q_i and Q_j $\geq \delta$ for all $i \neq j$ with respect to some Kähler metric of M.

(4.5) *Definition.* Let S' be a subset of S. The holomorphic subfamily $\mathcal{F}' = \{\Gamma_s\}_{s \in S}$ of $\mathcal{F}$ is said to a *strictly smaller open* subfamily if S' is a relatively compact connected open subset of S.

(4.6) *Definition.* Suppose we have a finite collection of smoothly simultaneously uniformizable families $\mathcal{F}_\mu$ of holomorphic curves in M with base points $(1 \leq \mu \leq p)$. Suppose each family $\mathcal{F}_\mu$ contains a strictly smaller open subfamily $\mathcal{F}'_\mu$. Let the symbols S_μ, S'_μ, Γ^μ_s, O^μ_s carry the meanings analogous to those of the corresponding symbols without the subscript or superscript μ. Let A_μ be the curve volume of $\mathcal{F}_\mu$ and assume that the orbit cardinality of $\mathcal{F}_\mu$ is at least N_μ. Let P and Q be two points of M. We say that Q *is linked to* P *via* the collection $\{\mathcal{F}'_\mu\}$ of families of curves if there exist $\mu_1, \cdots, \mu_k$ and there exists $s_\upsilon \in S_{\mu_\upsilon}$ for $1 \leq \upsilon \leq k$ such that $Q = O^{\mu_1}_{s_1}$ and $O^{\mu_{\upsilon+1}}_{s_{\upsilon+1}} \in \Gamma^{\mu_\upsilon}_{s_\upsilon}$ for $1 \leq \upsilon < k$ and $P \in \Gamma^{\mu_k}_{s_k}$. Let γ be a nonnegative number. By the *γ-linking constant from* Q *to* P we mean the infimum of

$$-c_1 \cdots c_k + \Sigma^k_{\lambda=1} d_\lambda (\Pi^{k+1}_{\upsilon=\lambda+1} c_\upsilon) - \frac{1}{m}$$

over all such choices of $\mu_1, \cdots, \mu_k$ and $s_\upsilon \in S_{\mu_\upsilon}$ $(1 \leq \upsilon \leq k)$ where $c_\upsilon = (1 - \frac{A_{\mu_\upsilon} \gamma}{2\pi N_{\mu_\upsilon}})^{-1}$ and $d_\upsilon = c_\upsilon \frac{A_{\mu_\upsilon}(\gamma+1)}{2\pi N_{\mu_\upsilon}}$ $(1 \leq \upsilon \leq k)$ and $c_{k+1} = 1$ (when all the c_υ's are positive).

The result on the existence of Kähler-Einstein metrics for manifolds with suitable finite symmetry is the following.

(4.7) *Theorem*. Suppose M is a compact complex manifold with a Kähler metric $h_{i\bar{j}}$ in its anticanonical class. Let G be a finite subgroup of the automorphism group of M. Suppose ψ is a smooth real-valued function on M such that with the conformal factor $e^{-\psi}$ the bisectional curvature of $h_{i\bar{j}}$ for two orthonormal vectors is bounded from below by some real number $-\kappa$. Let γ be a nonnegative number $> \kappa$. Suppose $\{\mathcal{F}_\mu\}$ is a finite collection of smoothly simultaneously uniformizable holomorphic family of complex curves in M with base points and each family $\mathcal{F}_\mu$ contains a strictly smaller open subfamily $\mathcal{F}'_\mu$. Let ρ and ϵ be two positive numbers. Suppose for every point Q of M there exists an open ball B_Q of radius ρ in M (with respect to $h_{i\bar{j}}$) such that Q is linked to every point in B_Q via $\{\mathcal{F}'_\mu\}$ with the γ-linking constant $< -\epsilon$. Then there exists a Kähler-Einstein metric on M. Moreover, the Kähler-Einstein metric can be obtained by solving by the continuity method the Monge-Ampère equation (2.1.1) with both $g_{i\bar{j}}$ and F invariant under the action of G.

§5. *Applications*.

This method can be applied to the Fermat cubic surface to give us a Kähler-Einstein metric on it. It can also be applied to higher dimensional Fermat hypersurfaces. With a rather minor modification it can be applied to give us a Kähler-Einstein metric on the surface obtained from $\mathbb{P}_2$ by blowing up three points. The modification involves using the torus group action and an additional step involving the mean value property of harmonic functions on domains in $\mathbb{C}$.

REFERENCES

[A1] T. Aubin, Métriques riemanniennes et courbure. *J. Diff. Geom.* **4** (1970), 383-424.

[A2] T. Aubin, Equations du type de Monge-Ampère sur les variétés kähleriennes compacts, *C.R. Acad. Sci. Paris* **283** (1976), 119-121.

[A3] T. Aubin, Equations du type Monge-Ampère sur les variétés kähleriennes compacts, *Bull. Sc. Math.* **102** (1978), 63-95.

[A4] T. Aubin, *Non-linear Analysis on Manifolds*, Monge-Ampère Equations. Springer-Verlag, New-York, 1982.

[A5] T. Aubin, Réduction du cas positif de l'équation de Monge-Ampère sur les variétés Kähleriennes compactes a la démonstration d'un inégalité, *J. Funct. Anal.* **57** (1984), 143-153.

[A-D] G. Averous and A. Deschamps, Estimées uniformes des solutions. In: *Premiere Classe de Chern et Courbure de Ricci: Preuve de la Conjecture de Calabi*, ed. J.-P. Bourguignon, Astérisque 58, Soc. Math. France 1978, pp.103-112.

[B-M] S. Bando and T. Mabuchi, Uniqueness of Einstein Kähler metrics modulo connected group actions, *Algebraic Geometry*, Sendai, 1985, Advanced Studies in Pure Matyh., Kinokuniya, Tokyo and North-Holland, Amsterdam, New York, Oxford.

[B-B-G] P. Berard, G. Besson, and S. Gallot, Sur une inégalité isopérimétrique qui généralise celle de Paul Lévy-Gromov, *Invent. Math.* **80** (1985), 295-308.

[Bi] E. Bishop, Conditions for the analyticity of certain sets, *Michigan Math. J.* **11** (1964), 289-304.

[B-C] R. L. Bishop and R. J. Crittenden, *Geometry of Manifolds*, Academic Press, New York-London, 1964.

[Bu] N. P. Buchdahl, Hermitian-Einstein connections and stable vector bundles over compact complex surfaces, Preprint of Max-Planck Institute, Bonn, 1986.

[C1] E. Calabi, The variation of Kähler metrics I: The structure of the space; II: A minimum problem, *Amer. Math. Soc. Bull.* **60** (1954), Abstract Nos. 293-294, p.168.

[C2] E. Calabi, The space of Kähler metrics, *Proc. Internat. Congress Math.* Amsterdam, 1954, Vol. 2, pp.206-207.

[C3] E. Calabi, On Kähler manifolds with vanishing canonical class, *Algebraic Geometry and Topology, A Symposium in Honor of S. Lefschetz*, Princeton Univ. Press, Princeton, 1955, pp.78-89.

[C4] E. Calabi, Improper affine hyperspheres and generalization of a theorem of K. Jörgens, *Mich. Math. J.* 5 (1958), 105-126.

[C5] E. Calabi, Métriques kähleriennes et fibrés holomorphes, *Ann. Sci. Ec. Norm. Sup. Paris* 12 (1979), 269-294.

[Ch] J. Cheeger, A lower bound for the smallest eigenvalue of the Laplacian, in *Problems in Analysis (A Symposium in Honor of S. Bochner)*, Princeton University Press, Princeton, 1970, pp. 195-199.

[C-L] S.Y. Cheng and P. Li, Heat kernel estimates and lower bound of eigenvalues, *Comment. Math. Helvetici* 56 (1981), 327-338.

[Cr] C. B. Croke, Some isoperimetric inequalities and eigenvalue estimates, *Ann. Scient. Ec. Norm. Sup.* 13 (1980), 419-435.

[D1] S. K. Donaldson, A new proof of a theorem of Narasimhan and Seshadri, *J. Diff. Geom.* 18 (1983), 279-315.

[D2] S. K. Donaldson, Anti self-dual Yang-Mills connections over complex algebraic surfaces and stable vector bundles, *Proc. London Math. Soc.* (3) 50 (1985), 1-26.

[D3] S. K. Donaldson, Infinite determinants, stable bundles and curvature, Preprint, 1986.

[E1] L. C. Evans, Classical solutions of fully nonlinear, convex, second order elliptic equations. *Comm. Pure Appl. Math.* 25 (1982), 333-363.

[E2] L. C. Evans, Classical solutions of the Hamilton-Jacobi Bellman equation for uniformly elliptic operators. *Trans. Amer. Math. Soc.* 275 (1983), 245-255.

[F] A. Futaki, An obstruction to the existence of Einstein Kähler metics, *Invent. Math.* 73 (1983), 437-443.

[F-M] A. Futaki and S. Morita, Invariant polynomials of the automorphism group of a compact complex manifold, *J. Diff. Geom.* 21 (1985), 135-142.

[G1] S. Gallot, Inégalités isopérimétriques, courbure de Ricci et invariants géométriques I, II. *C.R. Acad. Sc. Paris*, 296 (1983), 333-336; 365-368.

[G2] S. Gallot, A Sobolev inequality and some geometric applications. In: *Spectra of Riemannian Manifolds*, Kaigai Publications, Tokyo, 1983, pp.45-55.

[G-T] D. Gilbarg and N. S. Trudinger, *Elliptic Partial Differential Equations of Second Order*, 2nd ed., Springer-Verlag, 1983.

[H-N] G. Harder and M. S. Narasimhan, On the cohomology groups of moduli spaces of vector bundles on curves, *Math. Ann.* **212** (1975), 215-248.

[K] J. L. Kazdan, A remark on the preceding paper of Yau, *Comm. Pure Appl. Math.* **31** (1978), 413-414.

[K-W] J. L. Kazdan and F. W. Warner, Curvature functions for compact 2-manifolds, *Ann. of Math.* **99** (1974), 14-47.

[Ko] S. Kobayashi, Curvature and stability of vector bundles, *Proc. Japan Acad. Ser. A, Math. Sci.* **9** (1982), 158-162.

[Li] A. Lichnerowicz, Sur les transformations analytiques des variétés kählériennes, *C. R. Acad. Sci. Paris* **244** (1957), 3011-3014.

[Lü1] Lübke, Chernklassen von Hermite-Einstein-Vektorbündeln, *Math. Ann.* **260** (1982), 133-141.

[Lü2] M. Lübke, Stability of Einstein-Hermitian vector bundles, *Manuscripta Math.* **42** (1983), 245-257.

[M] Y. Matsushima, Sur la structure du groupe d'homeomorphismes analytiques d'une certaine variété kählerienne, *Nagoya Math. J.* **11** (1957), 145-150.

[M-R1] V. B. Mehta and A. Ramanathan, Semistable sheaves on projective varieties and their restriction to curves, *Math. Ann.* **258** (1982), 213-224.

[M-R2] V. B. Mehta and A. Ramanathan, Restriction of stable sheaves and representations of the fundamental group, *Invent. Math.* **77** (1984), 163-172.

[Mo1] J. Moser, A new proof of de Giorgi's theorem concerning the regularity problem for elliptic differential equations, *Comm. Pure Appl. Math.* **13** (1960), 457-468.

[Mo2] J. Moser, On Harnack's theorem for elliptic differential equations, *Comm. Pure Appl. Math.* **14** (1961), 577-591.

[Mo3] J. Moser, A sharp form of an inequality of N. Trudinger, *Indiana Univ. Math. J.* **20** (1971), 1077-1092.

[N-S] M. S. Narasimhan and C. S. Seshadri, Stable and unitary vector bundles on compact Riemannian surfaces, *Ann. of Math.* **82** (1965), 540-567.

[S] L. A. Santalo, *Integral Geometry and Geometric Probability* (Encyclopedia of Mathematics and Its Applications), Addison-Wesley, London-Amsterdam-Don Mills, Ontario-Sidney-Tokyo, 1976.

[Se] C. S. Seshadri, Space of unitary vector bundles on a compact Riemann surface, *Ann. of Math.* **82** (1967), 303-336.

[Si] C. Simpson, Systems of Hodge Bundles, Harvard University Ph.D. dissertation under the direction of W. Schmid, in preparation.

[Siu1] Y.-T. Siu, Kähler-Einstein metrics for the case of positive first Chern class, *Proceedings of Conference on Geometric Theory of Several Complex Variables*, Maryland, April, 1986, Springer-Verlag Lecture Notes in Mathematics.

[Siu2] Y.-T. Siu, The existence of Kähler-Einstein metrics on manifolds with positive anticanonical line bundle and a suitable finite symmetry group, Preprint 1986.

[T] N. S. Trudinger, Fully nonlinear, uniformly elliptic equations under natural structure conditions, *Trans. Amer. Math. Soc.* **278** (1983), 751-769.

[U] K. Uhlenbeck, Connections with L^p bounds on curvature, *Comm. Math. Phys.* **83** (1982), 31-42.

[U-Y] K. Uhlenbeck and S. T. Yau, On the existence of Hermitian-Yang-Mills connections in stable vector bundles, *Comm. Pure Appl. Math.* **39** (1986), 257-293.

[Y1] S. T. Yau, On Calabi's conjecture and some new results in algebraic geometry, *Nat. Acad. Sci. U.S.A.* **74** (1977), 1798-1799.

[Y2] S. T. Yau, On the Ricci curvature of a compact Kähler manifold and the complex Monge-Ampère equation, I, *Comm. Pure Appl. Math.* **31** (1978), 339-411.

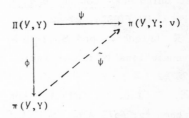